An Introduction to Marine

For David and Andrew, and Helen, Ruth and Anne

An Introduction to Marine Ecology

R. S. K. BARNES PhD, MA. Fellow of
St Catharine's College, Cambridge, and
Lecturer in Aquatic Ecology, University of Cambridge

R. N. HUGHES BSc, PhD, Lecturer in Ecology,
University College of North Wales

BLACKWELL SCIENTIFIC PUBLICATIONS

OXFORD LONDON
EDINBURGH BOSTON MELBOURNE

© 1982 by
Blackwell Scientific Publications
Editorial offices:
Osney Mead, Oxford, OX2 0EL
8 John Street, London
    WC1N 2ES
9 Forrest Road, Edinburgh
    EH1 2QH
52 Beacon Street, Boston
    Massachusetts 02108, USA
99, Barry Street
    Carlton, Victoria 3053, Australia

First published 1982

Set by Santype International Ltd
and printed and bound
in Great Britain by
Whitefriars Press Ltd, Tonbridge

DISTRIBUTORS

USA
    Blackwell Mosby Books
        Distributors
    11830 Westline Industrial Drive
    St Louis, Missouri 63141

Canada
    Blackwell Mosby Book
        Distributors
    120 Melford Drive, Scarborough
    Ontario, M1B 2X4

Australia
    Blackwell Scientific Book
        Distributors
    214 Berkeley Street, Carlton
        Victoria 3053

British Library
Cataloguing in Publication Data

Barnes, R.S.K.
    An introduction to marine ecology.
    1. Marine ecology
    I. Title   II. Hughes, R.N.
    574.5'2636   QH541.5S3

ISBN 0–632–00892–X

Contents

Preface

Several books have been written in which the words 'marine ecology' form part of the title and many more have been devoted to 'marine biology'. Why yet another? Once, when we knew very little of the lives and interactions of marine organisms, it was possible to outline the biology of the seas within one pair of covers; but it was possible only because of the extent of our ignorance. Who, during the same period, would have contemplated writing (or reading!) a book entitled 'terrestrial biology'? We now know much more of the biology of the sea, including its ecology, yet many introductory accounts still adopt the early descriptive approach concentrating on what organisms live where, how they are zoned in space, how they are adapted to their environment, etc. All worthy subjects, no doubt, but only a small part of 'ecology'. As university teachers faced with the task of providing undergraduates with their first taste of marine ecology, we have felt unable whole heartedly to commend any existing book as introductory or background reading to our courses. Available texts either suffered from the drawbacks mentioned above or were too specialized. Hence we set out to produce the sort of text which contained the information which we wished our students to know. To some extent the selection of topics and information is therefore a personal one, but we felt that there must be others besides ourselves faced with the same difficulties and who would find useful an introductory book concentrating on the trophic, competitive and environmental interactions of marine organisms and on the effects of these on the productivity, dynamics and structure of the marine system. Fortunately for us, Blackwell Scientific Publications agreed.

The book is therefore intended for students with some knowledge of ecology in general but about to venture for the first time into the *Mare Incognitum*. Each chapter covers a distinct process or subsystem of the ocean, and to that extent is complete within itself. Within this format, however, each possible subject-area is not covered in equal depth in each chapter: some processes operating throughout the sea are mainly described in relation to a single habitat type. Our knowledge of the effects of competition and predation on marine communities, for example, have largely been derived from work on rocky shores, and the following text cannot but reflect this.

It follows from the paragraphs above that we feel that the approach and synthesis presented here are different from those obtainable elsewhere—not least in our attempt to break away from 'classical marine biology'—and so we would like to hope that the book will also appeal to a wider audience as an introduction to

Preface

this rapidly growing subject.

Both authors accept responsibility for all the material in the book: each amended the labours of the other. Nevertheless, RSKB was mainly responsible for Chapters 1, 2, 3, 4, 7, 8, 11 and 12 and RNH for Chapters 5, 6, 9 and 10. Several people gave generously of their time and knowledge to save us from sins of omission and commission: in no way are they responsible for any errors which remain. Robert Campbell of Blackwell Scientific Publications was involved in the preparation of the book from the moment of its inception; we are sincerely grateful for his copious help, constructive criticism and advice. The authors are also delighted to record their appreciation of the labours of Jan Parr who drew the vignettes for many of the figures and who drew the illustrations of whole organisms for this book. The index was prepared by Hilary Barnes and Jean Hardy.

R. S. K. Barnes
Cambridge

R. N. Hughes
Bangor

1 The Nature and Global Distribution of Marine Organisms, Habitats and Productivity

1.1 Introduction

At the most gross level of analysis, the solid surface of the globe is formed by two different types of crustal materials: thin, dense oceanic crust and thick, light continental crust. Both float on the denser, upper layers of the mantle—the oceanic crust as a thin skin and the continental crust as a large lump—and both move in response to convection currents in that mantle. The oceanic crust is geologically young and is created continually along the mid-oceanic ridges; it then moves away from the ridges and is eventually resorbed into the mantle beneath the oceanic trenches. The continental crust, in contrast, is much older and it floats above, but is moved by, this sea floor spreading.

The existence of these two forms of crust is reflected in the earth's surface relief. Most of the ocean bed is a level expanse of sediment (with slopes of less than 1 in 1100) lying 3–4000 m below sea-level. From this plain, the huge continental blocks rise steeply, with an average upward slope of some 1 in 14 but in some areas with an average slope of more than 1 in 3, up to a depth of between 20 and 500 m (with an average of 130 m) below sea-level. At this point, representing the angle between the side and top of the continental mass, the gradient usually changes dramatically, falling to about 1 in 600. The average height of the top surface of the continent is less than 1000 m above this point.

Thus, ignoring the water for a moment, we have a scenario of a level ocean bed from which arise sheer-sided, topped or gently domed blocks averaging just over 5300 m in height. Also arising from our level plain would be volcanoes and the mid-oceanic ridges. Of course this is a considerable over-simplification, not least because the movement of sediment from the continents to the oceans tends to blur the starkness of the relief. Major rivers, when they discharge into the sea, do not lose their integrity but flow in submarine canyons scoured out by sediment-laden water. The 'river water' is no longer fresh but a dense suspension of sediment in sea-water and this descends the sides of the continental mass, the sediments being discharged into the angle between that mass and the oceanic bed. Great fans of sediment may extend out up to 600 km from the continental base, resting at an average slope of about 1 in 60 and forming ramps leading from the ocean floor to the continental sides.

This surface topography spans a height of almost 20 km (from the highest point on any continental block to the lowest point in an oceanic trench) but nevertheless this is an insignificant fraction of the earth's radius (0.3%). If, arbitrarily, we set the base of the rising continental blocks at a depth of 2000 m, then they would occupy 41% of the earth's surface. This is not to say that the land occupies that proportion of the surface, however, since as we have seen the sides and parts of the rims of the continents are below sea-level. Today, some 73% of the continental surface area projects above the waves and so the sea in total accounts for 70% of the earth's 510×10^6 km^2 surface—59% being that covering the ocean floor plus 11% over the submerged continental margins. The surface area of the continental masses appears more or less constant and so, therefore, is that of the ocean floor. However, the area of continental margin beneath the sea is, because of its shallow slope, subject to marked variation dependent on sea-level. A 100 m decrease in sea-level (such as might occur during the next glacial phase) would decrease the 11% of today to around 7%, whilst a 100 m rise in sea-level (which might result from the melting of all the earth's ice) would increase it to nearly 20%. Changes of this magnitude have occurred in the past and have greatly affected the abundance and diversity of the shallow marine fauna.

The sea therefore covers the largest portion of the earth's surface and it is even more important a habitat in terms of the total volume of the earth regularly inhabited by living organisms. On land the inhabited zone usually extends only a few tens of metres above the ground and a metre or so below it; the oceans are inhabited from their surface right down to their greatest depths (in excess of 11 000 m): the sea therefore provides 99% of the living space on our planet. Although the largest, it is also the least known and least knowable portion, particularly with regard to its biology.

Sir Alister Hardy likened our attempts to investigate the ocean to a person in a hot-air balloon slowly drifting over a land hidden from view by dense fog. Every so often, the balloonist would let down a bucket on the end of a large rope, let it drag along the ground for a while, and then, after pulling it up, examine the contents. What sort of an impression and understanding of terrestrial biology would an observer gain using such a technique? We have perhaps progressed beyond this limitation—but not very far. Maintenance of a research ship at sea is also very much more expensive than operating a hot-air balloon. We know most about life in shallow, coastal waters and about the relatively slow-moving and small to medium-sized organisms; we know least of life in the depths and of the smallest and largest, fastest-moving species. The reason for the depth limitation is self-evident. Note also, however, that fish population densities at depth are low (see

p. 293)—1 × 1000 m³ for example—and this imposes severe sampling problems. That relating to size of organism is not immediately obvious and is often overlooked. Most ecological information is still obtained from the sea by use of nets or by washing samples through a sieve. Neither nets nor sieves can be used to retain the smallest or most delicate organisms; they either pass through or are fragmented beyond recognition or study. Small water or sediment samples have to be taken *in situ* with consequent problems of sampling accuracy and adequacy, and with respect to bacteria, problems of changes in the relative proportion of the individuals and species originally caught after culture. Neither can nets capture large, fast-swimming species: they simply avoid the net. Our knowledge of the largest squids, for example, is entirely derived from the occasional specimen cast up on a beach (and examined by a biologist before scavengers and decay render it useless) and from the hard-parts (beaks) recovered from the whales which feed on them. Yet *Architeuthis* the giant squid may attain a total length of 17 m (and over 30 m has been claimed). Some whales (e.g. *Mesoplodon* and *Stenella* spp.) are also known only from a few specimens or fragments stranded on the shore, never having been seen alive. There is no scientific or probablistic reasons why Heuvelmans' (1968) thesis that several huge and as yet unknown fish and mammals occur in the oceans should not eventually be found to be correct.

There are therefore many areas of complete and almost complete ignorance, and there are several areas of controversy and doubt; but there is much that we do know and even more of which we are fairly certain. The following pages set out to introduce the reader to this body of knowledge and to present what currently appear to be the outlines of the ecology of the seas.

1.2 The nature of the ocean

We have already considered the basic shape of the crustal container housing the world's ocean and we must now put rather more flesh on these bones and describe those aspects of oceanic structure and those properties of sea-water that have a particular bearing on marine ecology. The study of marine science in general—oceanography—is of course a large field embracing physics, chemistry, geology and several other disciplines besides biology; here we must be very selective and many only marginally-relevant topics cannot be covered. The reader is referred to Wright (1977–1978) for additional information.

As we will see several times later, marine organisms appear

particularly to respond to and reflect three all-important environmental gradients: the latitudinal gradient in magnitude and seasonality of solar radiation from the poles to the equator (which will be deferred for detailed consideration to pp. 36–41); the depth gradient from the sea surface to the abyssal sea bed; and the coastal to open water gradient which often coincides with that in respect of depth. In fact all three are interlinked and are often superimposed.

The most straightforward of the three is the depth gradient. Although a host of terms have been coined for specific sections of the 0–11 000 m gradient, the most important distinction is between the uppermost few metres of the water column which can be illuminated by sunlight and the remaining 97.5% which cannot. Light is exploited for different purposes by different organisms, and different intensities will limit different processes. Therefore 'illuminated' must be qualified by reference to the process concerned. The light intensity at the sea surface also varies in regular diurnal and seasonal patterns and in relation to cloud cover, and hence any illuminated zone will vary with the light intensity and with the translucency of the water. Much of the incident light is scattered at the surface and of that which does penetrate this barrier, most is very quickly absorbed so that light intensity decreases logarithmically with depth. In the Sargasso Sea, for example, where the water is particularly translucent and light penetration is greatest, only a maximum of 1% of the red light penetrating the surface remains by the 55 m depth, only 1% of the yellow-green and violet light by 95 m, and only 1% of the blue by 150 m.

In order to photosynthesize, plants require light (particularly of the shorter wave lengths) and one can calculate the depth down to which the light is sufficient to permit their growth. In the most translucent, oceanic water and under conditions of full sunlight, the limiting depth for photosynthetic production is of the order of 250 m; in clear, coastal waters this reduces to about 50 m; and in highly turbid waters it is to be measured only in centimetres. Clearly, therefore, all primary fixation of organic compounds by photosynthetic organisms must be a phenomenon confined to the surface waters. The depth zone in which this is possible, the 'photic' (or 'euphotic') zone, averages some 30 m deep in coastal waters and some 150 m in the open ocean; the remainder of the depth gradient (and at night the whole ocean) is 'aphotic'. Below about 1250 m, there is insufficient sunlight for any biological process and hence, except for light produced by the organisms themselves, the ocean is thereafter lightless.

Light is one form of energy arriving from the sun; the second of great ecological consequence is heat. It is no accident that the element of a domestic kettle is situated at the bottom of the water mass enclosed within this heating appliance. As the water in contact with the element is warmed so it becomes less dense and rises, thereby allowing more, cooler water to replace it and be heated in turn. The heating process operates on convection currents which would not form if the element was positioned near the surface of the water mass. But, discounting geothermal sources, this is precisely the situation with respect to the source of heat and the oceans. The surface waters of the sea receive heat from the sun; therefore they become less dense and float at the surface: therefore they receive yet more heat: and so on. The end result is a body of hot, less dense water floating on top of a much larger mass of cold, dense water; the interface between the two, or more strictly the zone of rapid change in water temperature (Fig. 1.1), is termed the 'thermocline'. As with the photic zone, the position and magnitude of the thermocline are variable, but as water has a high specific heat it can absorb much heat with relatively little change in temperature and it will retain its heat for a long time in the presence of a temperature gradient. Diurnal changes in temperature are confined to the uppermost few metres and even there are rarely more than 0.3°C in the open ocean or more than 3°C in coastal areas.

The thermocline is therefore a feature of the upper 1000 m, below which the temperature of the ocean falls from a maximum of 5°C down to between 0.5 and 2.0°C. In contrast, at the surface, temperature may vary from −2°C to more than 28°C

Fig. 1.1 Characteristic temperature profiles at different latitudes in the open ocean. (The presence of cold, low-salinity water near the surface in polar latitudes disturbs the otherwise vertical profile there.) (After Wright 1977–78.)

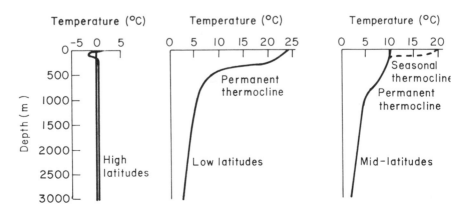

dependent on latitude (in contrast to fresh water, the density of sea-water decreases uniformly with a decrease in temperature down to near $-2°C$). Thermoclines are permanent features of the oceanic depth gradient in all but the highest latitudes, their magnitude depending on the temperature differential between surface and bottom waters. In regions experiencing an alternation of hot and cold seasons, a marked, though shallow and temporary, seasonal thermocline is superimposed on the relatively weak, permanent thermocline during the hot season (Fig. 1.1). The importance of this surface water/bottom water density difference is that it produces a barrier to mixing of these two water masses. Those dissolved substances taken out of the water in the photic zone and incorporated into living tissue which sink through the thermocline (as a result of gravitational forces) cannot be replaced by local mixing. Waters above a thermocline may therefore become depleted in these essential dissolved nutrients whilst the bottom waters hold large, untappable stocks (Fig. 1.2).

The third and final feature associated only with the surface

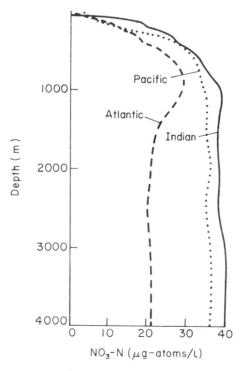

Fig. 1.2 The vertical distribution of nitrate in tropical and subtropical regions of the Atlantic, Indian and Pacific Oceans. (After Sverdrup *et al.* 1942.)

layers of the sea which must be mentioned here is wind-induced mixing. Winds blowing over the surface of the ocean impart some of their energy to the water, causing waves to form and inducing turbulent mixing of the surface layers down to maximum depths in the order of 200 m. This potential zone of mixing is therefore within the same depth range as the potential photic zone and the temporary seasonal thermocline of temperate latitudes; the precise relationships between these three depths at any one time are of great importance with respect to the potentiality of photosynthetic production. If, for example, mixing extends well below the photic zone, photosynthetic organisms may spend much more time below their threshold light intensity than above it and be unable to achieve sufficient production to balance their own energetic requirements (see pp. 45–9). Moreover, if mixing does not extend down to the thermocline, then photosynthetic production may also be reduced to low levels because of exhaustion of the nutrient supplies (Fig. 1.2).

The second major gradient is that stretching outwards from the coast into the open ocean, and it also involves variation in nutrients, depth and mixing. Several important subdivisions of the marine habitat can be made on this basis:

1. The immediately coastal or 'littoral' region from the upper limit of sea-water cover down to some 30 m depth.
2. The areas of submerged continental margins—the so-called 'neritic' water and the underlying 'continental shelf'.
3. The rapidly descending sides of the continental masses—the 'continental slope' with the more gently sloping 'continental rise' at the base of the slope.
4. The oceanic floor, usually termed the 'abyssal plain'.
5. The mid-oceanic ridges—vast mountain chains rising from the abyssal plain to within 2000 m or so of the surface (and occasionally breaking surface in the form of mid-oceanic islands).
6. The 'hadal regions' of the deep-ocean trenches—chasms in the abyssal plain descending from 6000 m to, in several cases, below 10 000 m.

The waters cradled within the continental slopes and the deep ocean floor are differentiated from the coastal neritic waters by being termed 'oceanic' (see Fig. 1.3). For present purposes, the three basic sections of the coastal to open water gradient are: (a) the littoral; (b) the neritic/continental shelf; and (c) the oceanic/abyssal (the latter including the continental slopes and rises, the mid-oceanic ridges and the deep trenches).

The essential feature leading to the separation of the littoral zone as a distinct part of the marine ecosystem is the extreme

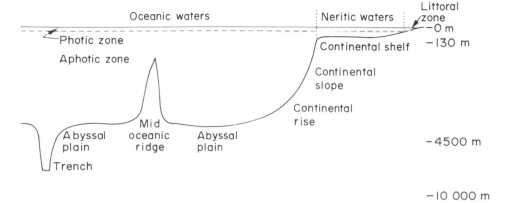

Fig. 1.3 Section of the ocean showing the major habitat subdivisions.

shallowness of the water. Light may penetrate to the sea bed and, indeed, in areas with tidal water fluctuations part of the bottom may become exposed temporarily to the air and receive solar radiation directly, diurnally or semi-diurnally. Plants associated with the sea bed can survive there and add their contribution to the total marine production. As we shall see later, the plants occupying this littoral fringe of insignificant area on a world basis may contribute quite markedly to this total. The transitional position of the littoral zone between land and sea has many other repercussions on the ecology of its characteristic organisms. Marine species, for example, may diminish interspecific competition by colonizing high intertidal levels; and the upper parts of marine shores may support luxuriant stands of semi-aquatic vegetation of basically terrestrial ancestry. Such salt-marshes, mangrove-swamps and, at slightly lower levels, sea-grass meadows (Chapter 4) also contribute significantly to essentially marine food webs.

Light more rarely penetrates to the bed of the somewhat deeper shelf seas (from 30 m to an average 130 m depth), and bottom-living, photosynthetic organisms are here much less significant. Nevertheless, shelf seas differ in many respects from open oceanic waters. Since wind action can keep surface waters mixed down to depths of some 200 m, shelf seas are well mixed and the impoverishment which can result from the loss through gravity of nutrients incorporated into tissue in the surface waters, is much less marked than in deeper systems. Coastal waters also receive the discharge of ground waters and rivers draining one hundred million square kilometres of land, and river water on average contains twice as much nutrient per unit volume as seawater (organic and inorganic materials dissolving in the water which percolates through rocks and the soil before being discharged by river and groundwater flow). For both these reasons (and in some areas also as a result of the upwelling considered on pp. 11–12), shelf seas are particularly productive.

8

Many bottom-living, coastal animals have evolved larval stages which swim for a time in the water, in part perhaps to exploit the productive, nutrient-enriched waters (pp. 239–49). Hence neritic waters may also be characterized by the abundance of these larvae. Once again therefore, although shelf seas only comprise some 3% of the ocean's area, they contribute much more than their proportional share to its total productivity. They are also locally very important and extend out from the coast for distances of up to 1500 km: such seas as the North and Baltic, Yellow and East China, Chukchi and Bering, Hudson Bay, much of the South China, Java, Arafura and Timor and even the Arctic Ocean are shelf seas.

Most of the world's ocean is the open sea cradled between the continental masses (Table 1.1). From what has been said above it

Table 1.1   Percentages of the world ocean comprised by various habitat types and depth intervals.

A. Habitat types	
Littoral zone	negligible
Continental shelf	3
Oceanic area	97
Continental slopes	12
Continental rises	5
Mid-oceanic ridges, mountains, etc.	36
Abyssal plain	42
Ocean trenches	2
B. Depth intervals	
0–1000 m	12
1–2000 m	4
2–3000 m	7
3–4000 m	20
4–5000 m	33
5–6000 m	23
> 6000 m	2

will be apparent that the surface waters of this region are relatively nutrient-poor, stable and overlie the cold, dark ocean depths. A marked change in the sediments also occurs. On the continental shelf, the sediments are predominantly sands, silts and clays of terrestrial origin (material eroded from the land by wave action, discharged by rivers or deposited by glaciers), locally with high percentages of the pulverized remains of molluscs, corals, etc., and with exposed bedrock in areas of rapid water movement. Over much of the abyssal plain, calcareous oozes formed from the skeletal remains of minute protists suspended in the water mass (e.g. coccolithophores and foraminiferans) dominate down to depths of 4500 m. The solubility of the carbonate ion ($CO_3^=$) varies with temperature and pressure, and below about 5000 m calcium carbonate goes into solution; hence below this depth there are no calcareous oozes. Other minute protists of the water column possess siliceous hard-parts (e.g. diatoms and radiolarians) and oozes formed largely of their remains are locally common in high latitudes, especially around Antarctica, and are also an important deep-sea sediment in low latitudes between depths of 4000 and 6000 m. In the dee-pest areas of the ocean (below 6000 m) the dominant sediment is a fine, inert red clay (Fig. 1.4). The open ocean as a whole is very poorly known; only of the surface photic zone do we have a reas-onable biological understanding.

Thus far we have neglected the properties of the water itself and to these we now turn. The water forming the world ocean is, of course, salty. In fact it approximates a 3.5% —or 35‰—solution (by weight) in which the dominant anion is chloride (19‰ by weight, comprising 86.8% of the total anions) and the dominant cation is sodium (10.5‰ by weight and 83.6% of the total cations). The concentration of salts in sea-water is remarkably constant; below 1000 m depth it varies only between 34.5‰ and 35.2‰, although clearly surface waters will be diluted

Fig. 1.4 Distribution of ocean-floor sediments. (After Wright 1977–78.)

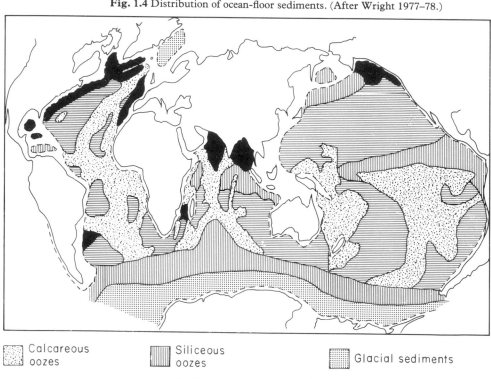

Calcareous oozes	Siliceous oozes	Glacial sediments
Red clay	Sediments derived from the adjacent land	

10

in areas of freshwater discharge and concentrated in regions with a marked excess of evaporation over precipitation—to over 36‰ in the tropical open oceans and to even higher values in semi-enclosed areas such as the Red Sea.

With the exception of the nutrient salts required by photosynthetic organisms, however, the chemical composition of sea-water and variation in this composition appear to play a very small role in marine ecology. Indeed, it has been argued (Barnes & Mann 1980) that the basic form taken by the ecology of aquatic systems is almost completely unrelated to the nature of the aquatic medium and is essentially similar in fresh, brackish and marine waters. Nevertheless, the abundance of such nutrients as nitrate and phosphate is often of very great significance. Both are only minor constituents of sea-water, nitrate averaging 0.5 $p/10^6$ and phosphate an order of magnitude less; both may reach potentially limiting concentrations in surface waters. Their distribution, abundance and flux, and the consequences for marine productivity, will reoccur frequently in the following pages.

Although, nutrients excepted, details of the composition of sea-water are of little ecological import (but of great interest in marine physiology), the movement of the water is critical and underlies many ecological processes and distributions, being particularly important in the circulation of nutrients and oxygen. Discounting tidally induced mixing which generally is of significance only in very shallow waters, the two main categories of large-scale water movement are density-driven currents and the various processes responsible for upwelling (and its converse, downwelling).

Upwelling, as its name suggests, is the movement of water from relatively deep in the ocean into the photic zone, i.e. movements parallel to the depth gradient and perpendicular to the surface. Its importance is that it is one of the few mechanisms by which the nutrient stocks of the aphotic regions can be introduced into the surface waters. Three processes may induce upwelling. First, deep currents when they meet such an obstacle as a mid-ocean ridge will be deflected upwards and may gush forth into surface waters. Secondly, when two contiguous water masses are moved apart, as for example when water immediately to the north of the equator moves northwards under the influence of Coriolis force and similarly water to the south of the equator is moved southwards, a 'hole' is left between them and water upwells to fill it. The depth from which water upwells is then dependent on the quantities of surface water moved laterally and on their current velocity. This is upwelling due to areas of divergence. Thirdly, and generally most importantly, when water is driven away from a coastline by wind action, an equivalent conceptual 'hole' is left which is filled by upwelling. In areas in which the continental shelf is of little extent, the upwelled water must be sucked straight from relatively deep zones (Fig. 1.5). We will look at the distribution of centres of upwelling below (pp. 38–9). Conversely, of course, areas of downwelling will occur in regions of convergence of water masses and when water is driven onshore: in each case, the water has nowhere to go but down, and thereby surface waters are entrained into the depths.

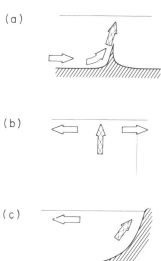

(a)

(b)

(c)

Fig. 1.5 Upwelling mechanisms consequent on: (a) an underwater ridge; (b) divergent surface currents; and (c) the movement of water away from a coastline.

The main currents in surface oceanic waters are driven by the prevailing winds (Fig. 1.6) and move along paths determined by the topography of the ocean basin and by Coriolis force (Fig. 1.7). These play an important role in the distribution of surface-living organisms, but, as we noted earlier, wind action does not penetrate below some 200 m depth and these surface currents have little influence on the large-scale movement and mixing of sea-water. The major currents in the sea as a whole are caused by changes in the density of certain surface water masses; and the currents generated, because of the shape of the sea bed, then downwell and cause compensatory upwelling. The density of sea-

Fig. 1.6 The general distribution of winds across the world ocean in January and June.

Fig. 1.7 The general pattern of surface currents in the world ocean.

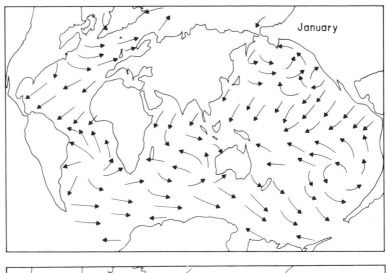

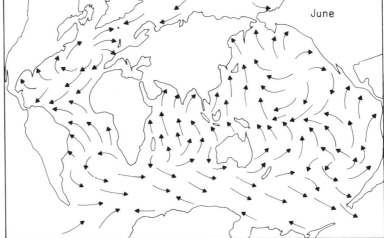

Fig. 1.6

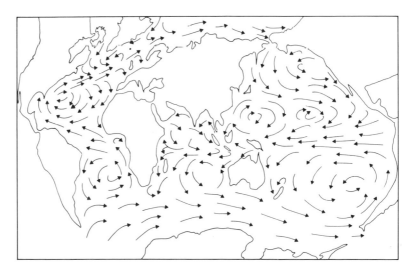

Fig. 1.7

water changes in relation to variation in temperature and/or salinity: cold sea-water is denser than warm, and full-strength sea-water is denser than that diluted by fresh water. Over the full ranges of temperature and salinity encountered in the oceans, salinity is a more potent factor in altering density than temperature: an increase in temperature from 7°C to 20°C decreases the density of sea-water by 0.002, a similar decrease to that achieved by diluting it from 36‰ to 33.5‰. In the oceans, changes in temperature and salinity usually occur together, and both can be illustrated by considering briefly the generation of the most important density-driven current in the ocean—that emanating from the south polar seas.

The Antarctic continent is glaciated, centred on the South Pole and surrounded by sea-water, and contact with ice will alter both the temperature and salinity of sea-water. The ice cools the water immediately adjacent to it, and being cooled the water is rendered more dense and hence it sinks. Of necessity, the sinking water must flow northwards as it descends the continental shelf and slope. During the Antarctic winter, fresh water also freezes out of the surrounding sea and increases its salinity, thereby further increasing its density. Water upwells 'to make good the loss'. Part is rendered more dense through decreased temperature and increased salinity and in sinking continues the downwelling process; part flows northwards at the surface, in the Antarctic summer aided by the addition of fresh water from melting ice. Therefore two currents are generated both flowing northwards in all latitudes, one moving along the sea bed, the other on the surface. The bottom current is by far the larger and more important: vast volumes of water are involved (twenty million cubic metres per second) and the current can be traced well north of the equator (to 40°N). A comparable system operates in the Arctic, and in the Atlantic a deep current centred at about 3000 m, which has entered this ocean near Greenland, can be traced to about 40°S (Fig. 1.8).

These currents effect one major difference between freshwater lakes and the oceans. In fresh waters, permanent thermoclines are usually associated with deoxygenated bottom waters. However, the density-induced currents of the oceans ensure that the oxygen dissolved in surface waters is transported to the deepest parts of the sea: with very few exceptions, sea-water is well oxygenated regardless of depth, and although sea-water can hold less oxygen, volume for volume, than fresh water, there is nearly always sufficient oxygen in sea-water to meet the respiratory requirements of its contained organisms. The exceptions are three-

Latitude

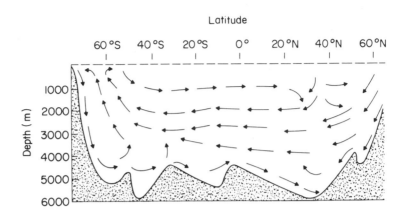

Fig. 1.8 Section of the Atlantic Ocean showing the deep currents. (After Wright 1977–78.)

fold. A few, largely land-locked seas, such as the Black Sea and the Gulf of California, are partially isolated from the larger, adjacent water masses by sills at their mouth, and, as in lakes, deoxygenation of their semi-stagnant, bottom waters can result (the addition of a pollutional oxygen demand to the Baltic Sea is having the same end result (see p. 300)). In a small number of well-defined areas, the rate of production of organic matter and the rate of its sedimentation towards the bottom exceed the rate at which the oxygen required in its decomposition can be supplied by deep currents, and again local temporary anoxia can occur, for example, off the coast of south-west Africa. Finally, a zone termed the 'oxygen minimum layer' occurs in the ocean, usually between depths of 400 m and 1000 m (Fig. 1.9). This arises because although oxygen is being removed from the water at all depths by respiration, it is replenished from two opposing directions, from the surface where the water is in contact with atmospheric oxygen and is usually supersaturated, and from the depths where the bottom currents described above continually import oxygen-laden surface water from elsewhere. Between the two will be a zone in which the oxygen removed from the water is maximal in relation to the rate of influx, and in this minimum layer concentrations may fall from the 4–6 mg l^{-1} norm down to less than 2 mg l^{-1}. It must be stressed, however, that these regions of oxygen deficiency are markedly atypical of the sea as a whole, and oxygen levels are very rarely a factor of relevance in marine ecology. (This has not always been so: life probably originated in anoxic seas and some have argued that the obligate anaerobes representing several metazoan phyla which occur today in the deep, anoxic regions of marine sediments are primitive survivors of an original metazoan adaptive radiation which occurred when the seas were still oxygen free. Even in the Cretaceous large areas of sea bed may have been anoxic.)

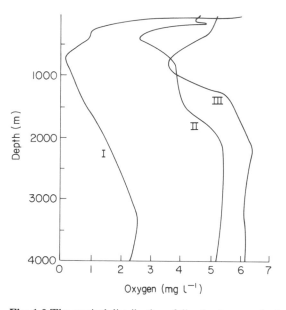

Fig. 1.9 The vertical distribution of dissolved oxygen in the ocean, showing the presence of an oxygen minimum layer to the south of California (I), in the eastern South Atlantic (II), and in the Gulf Stream (III). (After Friedrich 1965.)

1.3 The nature of marine organisms

Several popular definitions of ecology involve interactions between organisms and their environment, and having introduced the environment it is now time to introduce the organisms. This is not the place to give a systematic account of the thousands of marine species and their classification. Rather, this section will be ecologically based; nevertheless it is important to start with some idea of the phylogenetic diversity of marine life.

For many years it was customary to force all living organisms into one or other of the categories 'animal' and 'plant', and for botanists and zoologists to urge rival claims on groups which did not easily fit these concepts. More recently there has been an increasing tendency to adopt a more realistic series of categories (kingdoms): Monera (the prokaryotic bacteria and blue-green algae), Protista (single or non-cellular eukaryotes and their immediate relatives), Fungi (the 'higher' multicellular saprophytes), Plantae (the 'higher' green plants), and Animalia (metazoan animals) (see Whittaker 1969; House 1979). These make greater ecological—as well as phylogenetic—sense and will be used here. Representatives of all these kingdoms occur in the sea, although the Fungi and Plantae are mainly terrestrial. Opinions on the number of major groupings of organisms (phyla) to be included in these kingdoms vary; Whittaker (1969) lists 51 containing living representatives and in some more recent lists almost twice as many are recognized. In any event, excluding a few groups comprising solely parasites, all (or in some classifications all but one or two) of these phyla are represented in the sea or in the transitional coastal fringe. Marine examples of the majority of these phyla are illustrated in Sieburth (1979), Bold and Wynne (1978) and George and George (1979), and a recent review of their phylogenetic relationships is provided by House (1979).

In contrast, land and freshwater organisms represent a much smaller range of known phyla, classes and other higher taxonomic categories. This reflects the origin and early diversification of life in the sea and also the problems faced by organisms in any attempt to increase their range by adapting to the more rigorous, more changeable, and essentially non-aqueous environment of the land. Only the ascomycete and basidiomycete fungi, the tracheophyte plants and three groups of animals, the chelicerate and uniramian arthropods and the tetrapod chordates, have broken the barrier and become more characteristic of the terrestrial sphere than of the marine.

Yet when one looks at the relative numbers of individual species the picture is completely reversed. There are probably some five million described and undescribed species of living organisms and of these less than 250 000 are marine! Those few groups which have managed to leave the ancestral home of the sea, particularly the insects, have clearly diversified to an incredible extent. This tells us much about the process of speciation in general. Diversification requires the isolation of small populations by geographical barriers to dispersal. These isolating barriers are present *par excellence* on land, but the ocean is continuous and, relatively, stable and uniform. These conclusions are reinforced by the observation that a small minority of marine species (about 2%) inhabit the water mass itself; most species are associated with the more diverse and more fragmented bottom sediments.

Degree of diversification into species is one measure of success but it is not the only one nor, indeed, one of the better. Abundance of a group (whether measured as number of individuals, biomass or productivity) and length of time for which a basic body plan has been able to persist are other measures. In these terms, many marine groups are highly successful.

Regardless of their phylogenetic position, marine organisms can be placed in two large categories dependent on whether they live in the water mass (pelagic) or on or in the bottom sediments or rock (benthic). A minor, third category is required for those organisms that straddle the air–water interface (pleustic). These categories, although extremely useful, are by no means mutually exclusive or rigidly definable. We have already seen (p. 9), for example, that some species are benthic as adults but pelagic as larvae, and a number of pelagic organisms may spend much time resting on or feeding at the sediment–water interface (these are often termed bentho-pelagic).

If organisms are positively buoyant they will float at the air–water interface like corks. In some animals, sea-birds such as auks, petrels and shearwaters for example, flotation serves as a resting phase from which to depart either into the water for food or into the air for longer distance dispersal. In others, their association with the boundary layer is permanent. This may be achieved by the possession of a float, as in some coelenterates (the Portuguese man o'war, *Physalia*, and by-the-wind-sailor, *Velella*), crustaceans (the stalked barnacle *Lepas fascicularis*) and molluscs (the gastropod *Ianthina* which secretes a frothy raft and the opisthobranch *Glaucus* which has bubbles of air in its gut), or by utilizing the surface tension of the interface. The latter property is exploited by the sea-skater *Halobates*, a hemipteran insect which strides across the surface film in the same fashion as the more numerous pond-skaters of fresh waters (Fig. 1.10). Intuitively one might think that an existence tied to this interface, being distributed at the mercy of surface current and wind direction, and without any mechanism for changing depth, etc., would have more than its fair share of difficulties, and indeed the number of pleustic species is very small—all are largely confined to the tropics. Most are carnivores (in the broadest sense of the word), either passively dangling a food-catching apparatus down into the water or being dependent on other species of the pleuston.

The pelagic category is very much larger and it includes all those species of the water column which are distributed at the mercy of currents (the plankton), in that their powers of locomotion are insufficient to enable them to make headway against

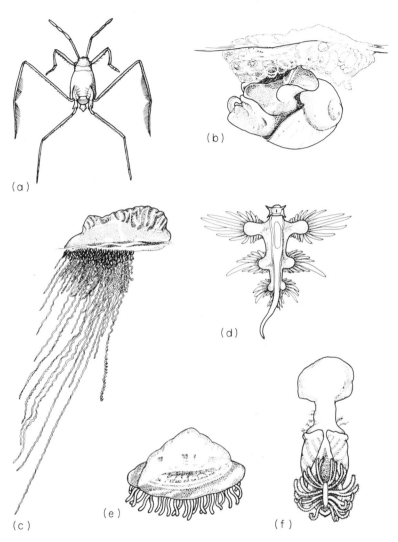

Fig. 1.10 Representative members of the pleuston: (a) *Halobates;* (b) *Ianthina;* (c) *Physalia;* (d) *Glaucus;* (e) *Velella;* and (f) *Lepas fascicularis.*

current action, and those, usually larger, species which can swim more powerfully (the nekton). Swimming ability is usually related to size and all gradations are found—from minute organisms without any means of propulsion to large animals capable of migrating from the Arctic to the Antarctic Ocean and back again at will. Like pelagic and benthic, therefore, plankton and nekton are mainly terms of convenience.

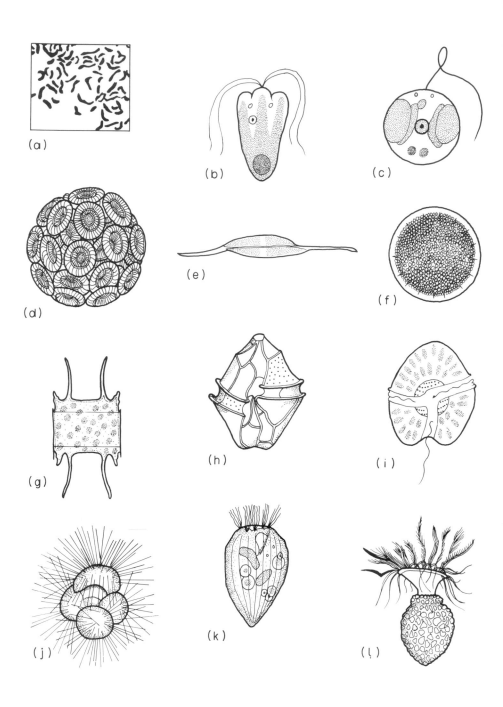

Fig. 1.11 Representative members of the ultra-, nano- and microplankton: (a) bacteria; (b and c) flagellates; (d) coccolithophore; (e – g) diatoms; (h and i) dinoflagellates; (j) foraminiferan; (k) ciliate, and (l) tintinnid.

The plankton suspended in sea-water range in size and phylogenetic position from bacteria of less than 1 μm in diameter to relatively large jellyfish exceeding 0.5 m across the umbrella (Table 1.2 and Figs. 1.11 and 1.12). Since our knowledge of them is derived from the use of nets and filters of standard pore size, it is usual to refer to different (logarithmic) size classes: ultraplankton (< 2 μm); nanoplankton (2–20 μm); microplankton (20–200 μm); macroplankton (200–2000 μm); and megaplankton (> 2000 μm).

Table 1.2 Systematic list of the main groups of planktonic organisms (parasites excluded).

Monerans	Bacteria
	Cyanophytes (blue-green algae)
Protists	Cryptophytes
	Dinophytes (dinoflagellates)
	Euglenophytes
	Chrysophytes
	Bacillariophytes (diatoms)
	Haptophytes (e.g. coccolithophores)
	Prasinophytes
	Xanthophytes
	Chlorophytes (green algae)
	Several groups of amoeboid and flagellate protists
	Foraminiferans
	Radiolarians
	Acantharians
	Ciliophorans (ciliates)
Animals	Medusozoan coelenterates (medusae, jellyfish, siphonophores, etc.)
	Ctenophores (sea gooseberries)
	Gastropod and cephalopod molluscs
	Polychaetes
	Ostracod, copepod and malacostracan crustaceans
	Chaetognaths (arrow-worms)
	Thaliacean and appendicularian tunicates
	Fish

(Together with the larvae of otherwise benthic groups)

This supplements and in part corresponds to a systematic classification in that the ultraplankton are chiefly bacteria, the nano- and microplankton are protists, and the macro- and megaplankton are animal. It is normal practise to refer collectively to the photosynthetic members of the nano- and microplankton as phytoplankton (and, together with their larger seaweed relatives of the littoral, as algae), and to the heterotrophic species of micro-, macro- and megaplankton as the zooplankton. The relative abundance of the different size groups in one water mass of

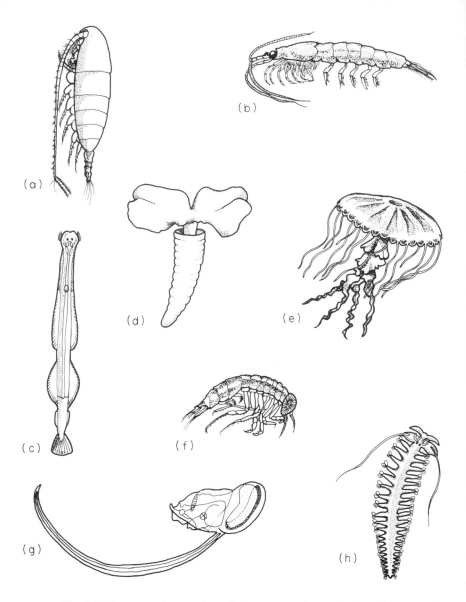

Fig. 1.12 Representative members of the macro- and megaplankton: (a) copepod; (b) euphausid; (c) chaetognath; (d) pteropod; (e) jellyfish; (f) hyperiid amphipod; (g) appendicularian; and (h) polychaete.

the south-western coast of Canada is shown by way of example in Fig. 1.13.

The plankton display a number of interesting features. It is evident that the phytoplankton, which, with the exception of the littoral zone, are the sole source of the food energy which fuels the entire oceanic food web, are between 2 and 200 μm in size. On land, the dominant photosynthetic organisms are many orders of magnitude larger, and are often larger than the herbivores which graze them. Why then are the marine autotrophs so small? There are several answers to the question. Plants are large

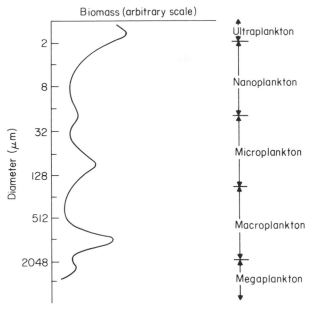

Fig. 1.13 The relative abundance of different size-classes of plankton off the south-western coast of Canada. (After Parsons *et al.* 1977.)

on land in part because they need supporting tissue—woody material enabling the photosynthetic apparatus to be raised nearer the light source than their competitors. Tree-shaped plants in the sea would need, on average, to be 4000 m tall, and for 3800 of those metres would have to grow without external supplies of energy: this strategy is clearly a non-starter. But why have the protists not produced a large, floating, multicellular organism which could intercept more light than smaller, potentially-competing species? One of the problems facing all species which need to remain in the photic zone is the gravitational attraction exerted on all living tissue, which, because of the included salts and organic molecules, is denser than seawater. Here being small helps. A particle 2 μm in diameter will sink only 10 cm through distilled water at 20°C in a time of over 8 h; whereas, under the same conditions, a 62 μm diameter particle will sink 10 cm in less than 30 seconds. Weight is proportional to volume, and frictional resistance to surface area, so small objects with a large surface area per unit volume will sink relatively slowly—small phytoplankton cells will effectively be suspended in the water and their rate of fall will be dwarfed by water movements resulting from wind-induced turbulence.

These turbulent water movements may carry many photosynthetic organisms out of the photic zone, however, and herbivores may consume many algae per unit time. Hence in order to make good these and other losses and to maintain populations in the photic zone, the phytoplankton must be able to multiply rapidly. Here again, small size is important; available energy must be devoted to multiplication rather than put into individual growth. The intrinsic rate of population growth, r_{max}, is in living organisms related to weight (Wg) by the expression, $r_{max} = 0.025$ $W^{-0.26}$. An organism 10 μm long has a generation time of one hour; one 10 m long has a generation time of ten years. A third and final consideration is nutrient uptake. Terrestrial plants take up nutrients through their root systems which have a very large surface area in relation to the plant's volume. Planktonic protists cannot afford to increase their mass, so in order to have a large enough surface area to take in sufficient nutrients to support their mass/volume-dependent metabolism they must be small. It is even generally true that within the size range of the phyto-plankton, the smallest species are characteristic of areas with a premium on efficient nutrient uptake and on multiplication to offset herbivore pressure.

Shape as well as size is relevant: phytoplankton are rarely spheres (see p. 226). Many have long spinous processes and others are completely spine shaped. This may help to slow the sinking rate even further, but more importantly it increases the effective size of an organism from the point of view of a potential consumer. A sphere of given volume is easier to handle and consume than a needle of equivalent volume.

The non-photosynthetic protists and animals of the plankton face exactly the same problems and have solved them in a similar fashion. Members of the macro- and megaplankton do possess locomotory systems more effective than flagella and cilia at enab-ling them to counteract gravity, but nevertheless they also pos-sess adaptations lowering their specific gravity. These include gelatinous tissue, the presence of oil droplets and the replacement of heavy bivalent cations by light monovalent ones. They too may also bear long spines or setae increasing their effective body size.

One problem characteristically facing the consumers within the plankton is that of finding more food when that in their immedi-ate vicinity has been exhausted. A terrestrial animal could merely walk or fly to another patch, i.e. change its location within the habitat, but since by definition plankton move with their environment and cannot overcome current action, this is not nearly so easily achieved. Most terrestrial animals, however, can only move in a horizontal plane, whilst the plankton have a large

vertical component in their habitat and this makes all the difference. If one is travelling in a fast moving ship or bus, one cannot
effectively alter one's movement with respect to the earth's surface by moving along the vehicle—its motion is much greater
than one's own—but it is a simple matter to move in a vertical
plane and climb from one deck or floor to another. In the sea,
different water masses at different depths move at different
speeds and in different directions (see e.g. Fig. 1.8), and these
different currents are analogous to the various decks of our hypothetical ship. By moving downwards a planktonic organism will
eventually enter a different current and on ascending back near
the surface again a new patch of surface water will be encountered (Fig. 1.14). In effect, the only way for a marine organism with limited powers of locomotion to change its horizontal

Fig. 1.14 Vertical migration between two water masses flowing in different
directions and/or at different speeds can effect a change in the surface water mass
inhabited by the migrating organism.

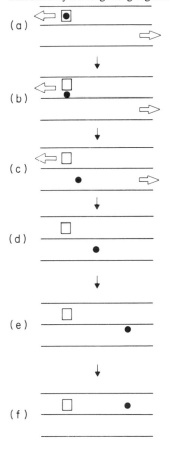

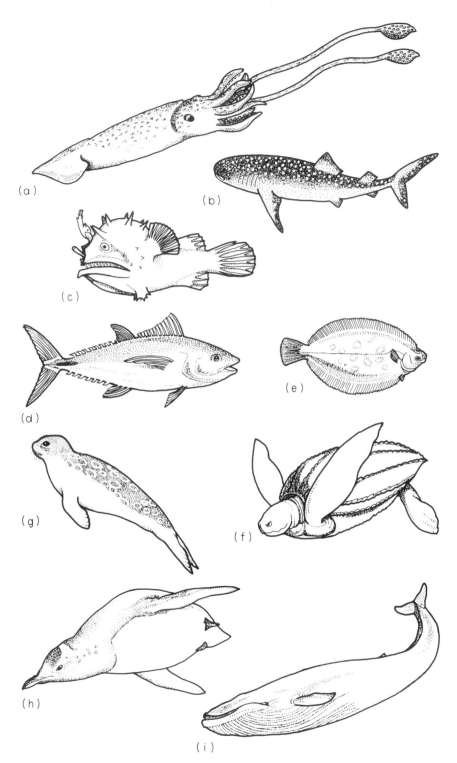

Fig. 1.15 Representative members of the nekton: (a) squid; (b) shark; (c) deep-sea fish; (d) tunny; (e) flat-fish; (f) turtle; (g) seal; (h) penguin; and (i) whale.

position in the photic zone is to move vertically: one migration downwards and then back up to the surface again (see pp. 67–70).

Plankton occur throughout the oceans from the uppermost few centimetres of the water layer (where a characteristic assemblage of species—the neuston—are permanently associated with the undersurface of the air–water interface; see David 1965) down to the abyssal depths. Greatest development, in terms of abundance and productivity, not surprisingly is located in the photic zone and there the plankton forms a self-sufficient community in all but complete nutrient regeneration. Algae fix carbon photosynthetically and are consumed by herbivores which in turn are taken by carnivorous zooplankton (both herbivores and carnivores releasing inorganic nutrients back into the water). The bacterioplankton subsist on soluble organic compounds released by the phytoplankton, on the faeces of the zooplankton and on dead organisms of all types. Bacteria are engulfed by the protistan zooplankton, and these protists, other bacteria, and the detrital substrate are eaten by omnivorous and detritus-feeding members of zooplankton. In addition to these permanently planktonic organisms (the 'holoplankton'), other species acting as consumers occur in the form of the larval stages (the 'meroplankton') of, for example, nektonic fish and, especially in neritic waters, of benthic invertebrates. Although self-sufficient, the plankton is not closed. Except in the littoral zone, it entirely supports the nekton and benthos as well.

The nekton (Fig. 1.15) are differentiated from the plankton only on the basis of swimming ability. The large cephalopod molluscs (squid), crustaceans (e.g. prawns) and vertebrates (fish, turtles, sea snakes, seals, sea-cows, whales, etc.) comprising this category of pelagic organisms are the terminal consumers of the sea, mostly being carnivorous although a few are herbivores and even fewer take detritus. Like most top predators, they need to range widely in search of food concentrations—and hence the swimming ability which characterizes them. Swimming is not only of use in moving relative to the earth's surface; as anyone who has watched fish in a river will know, a fish also needs to swim against a current to stay in the same place. In this manner nektonic organisms may establish territories in such areas as coral reefs and remain there against the water flow. Active swimming requires the development of muscular systems and often of relatively large size. The nekton are therefore those pelagic species with the greatest energy requirement simply to counteract gravity. Some fish and cephalopods have evolved gas-filled buoyancy

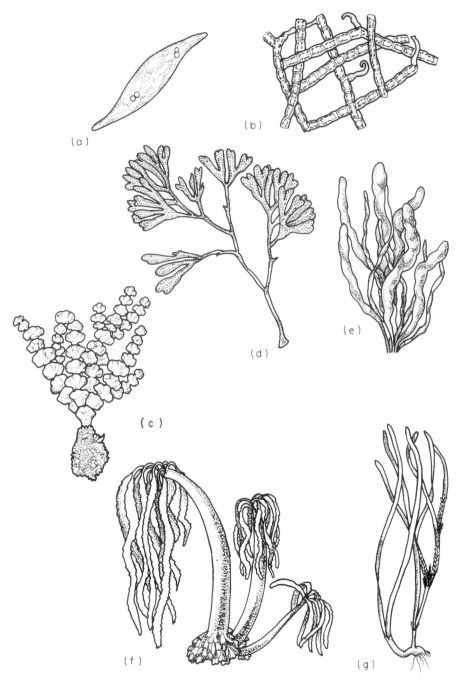

Fig. 1.16 Representative members of the phytobenthos: (a) benthic diatom; (b) mat-forming alga; (c) coralline alga; (d) wrack; (e) *Enteromorpha;* (f) the kelp *Postelsia;* and (g) sea-grass.

devices to achieve neutral density but a concomitant of this is a lessened ability to change depth in the water column rapidly. Nekton which chase prey actively have generally dispensed with buoyancy aids and maintain themselves in the water solely through constant locomotion. Others have abandoned a perpetually pelagic life and rest on the bottom between active forays for food or even whilst feeding (the bentho-pelagic groups).

The problem is most intense for the deep-sea fish which live in a habitat with low food availability. Sufficient energy to maintain an active swimming regime in order to achieve suspension is not available, yet they are pelagic feeders and must also maximize the chances of prey capture. The solution has been a bizarre suite of adaptations. Body tissue is largely reduced to a gelatinous consistency (so that a 1.5 m long fish may weigh only as many kilograms); this reduces both gravitational attraction and maintenance metabolic requirements. Prey is often lured towards the floating fish-trap by bioluminescent organs and, when in range, it is engulfed by a cavernous mouth, the predator retaining only sufficient muscular ability to make the short, sharp lunge required. Since prey density is low, response to the lure is infrequent, and any 'bite' must be seized almost regardless of size. Jaws can be dislocated and stomachs distended to accommodate prey items almost twice the length of the predator.

In spite of the efficiency of food uptake engendered by food shortage, some of the potential food available in the water column escapes utilization by pelagic consumers and settles on the sea bed, largely in the form of faecal pellets and other items low in nutritive value. This then forms the basic food resource of the benthos, and in general it will be the more abundant the smaller the distance between sea bed and photic zone (although anomalies may occur where deep water lies immediately adjacent to land and can receive an input of river-borne debris, etc.).

Except in the littoral zone, the benthos is lacking in photosynthetic organisms. Chemosynthetic bacteria (nitrifying, sulphur, hydrogen, methane, carbon monoxide, and iron bacteria) occur in benthic sediments, however, and can there fix carbon dioxide by oxidizing ammonia, hydrogen sulphide, methane, etc., in some cases using dissolved oxygen but in others by utilizing that bound in, for example, sulphates. In many cases, the inorganic substrates used are produced by the decomposition of organic matter produced earlier and elsewhere by photosynthesis, and organic compounds themselves may be used as carbon sources: in these cases, chemosynthesis is only regenerating organic matter. But some bacteria can produce organic materials quite independently of photosynthesis. Reduced sulphur compounds issuing from submarine volcanic vents —fumaroles —are used as energy sources by some sulphur bacteria in their fixation of dissolved carbon dioxide.

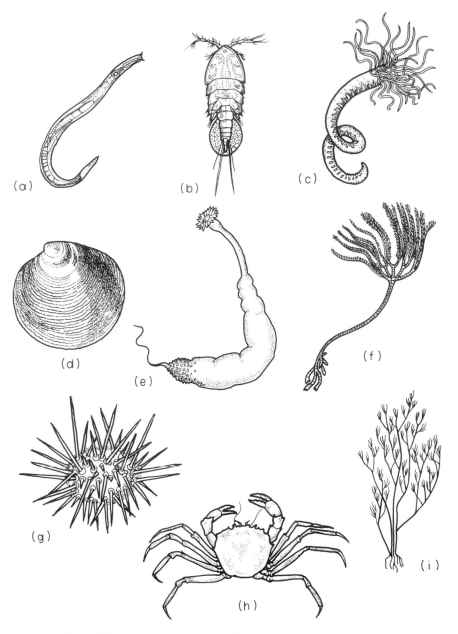

Fig. 1.17 Representative members of the zoobenthos: (a) nematode; (b) harpacticoid copepod; (c) polychaete; (d) bivalve; (e) sipunculan; (f) stalked crinoid; (g) sea-urchin; (h) crab; and (i) bryozoan.

As for the plankton, benthic organisms are isolated for study using pores of known size (in sieves and filters) and hence a size-based series of categories is also used: microbenthos ($< 100\ \mu$m); meiobenthos (100–$500\ \mu$m); and macrobenthos ($> 500\ \mu$m). The microbenthos comprises bacteria and protists; the meiobenthos, animals together with a few large protists (particularly foraminiferans); and the macrobenthos category is, except in the

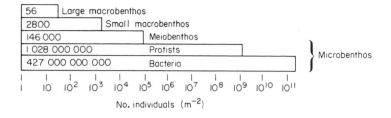

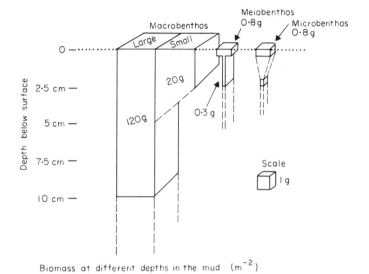

Biomass at different depths in the mud (m⁻²)
(a)

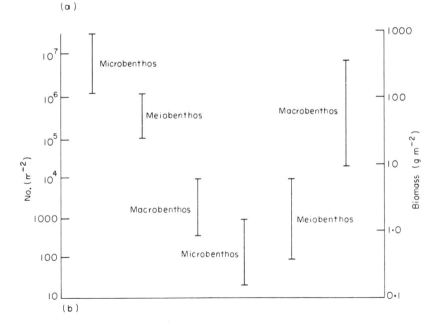

(b)

Fig. 1.18 The relative abundance, in terms of density and biomass, of different size-classes of benthos. (a) In mud off south-western England. (After Mare 1942.) (b) In sandy sediments of the continental shelf in general (excluding bacteria). (From data in Fenchel 1978.) In the deep sea, the biomass of meio- and macrobenthos are more nearly equal.

littoral zone, entirely animal (Figs. 1.16 and 1.17). Fig. 1.18 illustrates their relative abundance. The benthos includes a much more diverse collection of organisms than the plankton, especially of animals (Table 1.3) in which all phyla with marine examples are represented. A distinction between 'epifauna' and 'infauna' is almost universal, the epifauna living on the sediment–water interface rather as *Halobates* lives on the air–water interface, and the infauna living within the sediment itself. Comparably to the pleuston, however, several animals may span the interface by extending a feeding system into the overlying water whilst the rest of the body is hidden within the substratum.

Table 1.3 Systematic list of the main groups of benthic organisms (parasites excluded).

Monerans	Bacteria
	Cyanophytes* (blue-green algae)
	Bacillariophytes* (diatoms)
	Chlorophytes* (green algae)
	Euglenophytes*
Protists	Phaeophytes* (brown algae)
	Rhodophytes* (red algae)
	Several groups of amoeboid protists
	Foraminiferans
	Ciliophorans (ciliates)
Plants	Tracheophytes*
	Poriferans (sponges)
	Medusozoan and anthozoan coelenterates (hydroids, corals, etc.)
	Turbellarians (flatworms)
	Gnathostomulids
	Nemertines
	Gastrotrichs
	Kinorhynchs
	Nematodes
	Priapulids
	Entoprocts
	Molluscs (all groups)
	Sipunculans
	Echiurans
Animals	Polychaete and oligochaete annelids
	Pogonophorans
	Tardigrades
	Pycnogonid and merostomatan chelicerates (sea-spiders, king-crabs, etc.)
	Crustaceans (all groups)
	Phoronids
	Bryozoans
	Brachiopods
	Echinoderms
	Hemichordates
	Ascidian tunicates
	Cephalochordates
	Fish

* Groups confined to the littoral zone.

As one might expect, on hard substrata such as exposed bed-rock most organisms are epifaunal, whereas most animals are in-faunal in soft sediments; but in the deeper areas of the abyssal plain and in the ocean trenches, many species in groups which are characteristically infaunal are to be found on the surface of the ooze. This change of living station is probably to be associated with decrease in predation intensity. Several authors (e.g. Clark 1964; Mangum 1976) have advanced the thesis that the earliest meio- and macrobenthos were epifaunal and remained thus until the advent of large predatory species which could roam over the sediment surface. These predators exerted a strong selection pressure in favour of a life spent in hiding beneath the surface. In coarse sediments, meiobenthos can move between the grains without displacing them, but macrobenthos must burrow. Clark suggested that fluid-filled body cavities first evolved in this connection—as hydrostatic skeletons permitting burrowing—and Mangum argued that burrowing would have resulted in problems in obtaining oxygen and that the 'invention' of respiratory pig-ments by the annelids went hand-in-hand with the invasion of the sediments. As the density of potential prey falls in the deep sea (pp. 196–9), so life as a predator becomes progressively more difficult and this mode of nutrition is poorly represented below 6000 m (members of otherwise largely predatory groups, e.g. the starfish, turning to ooze-eating instead). The absence or scarcity of predators has then permitted animals to return to the ancient epifaunal existence.

Over the whole of the continental shelf and abyssal plain, two feeding types predominate: suspension feeding on particles car-ried by or settling out of the water; and deposit feeding on ma-terial on or in the sediment. In both cases, the ultimate food or-ganisms are probably bacteria subsisting on the rain of refractory debris from the pelagic system. The bacteria may be captured and consumed directly or only after passage along a bacteria ⟶ protist ⟶ meiofauna chain, but below about 30 m (or whatever the depth of the photic zone may be) the benthos will be depen-dent on bacterial activity. Deposit feeding is viable on all types of sediment, although large quantities may have to be consumed, but suspension feeding will only be successful if there are suf-ficient particles of potential food suspended in the water. There-fore, suspension feeders are most characteristic of shallow regions and of areas of relatively rapid water movement into which an animal may extend a passive filter and at intervals wipe off parti-cles retained.

The benthos of the littoral zone has been reserved for comment separate from that of the shelf and abyssal plain, because it is the zone to which benthic photosynthesis is confined and in which the latter may exceed that in the water column. Three categories of photosynthetic organisms occur: (1) protists essentially similar to those of phytoplankton but here associated with soft sediments, symbiotic within littoral animals such as reef corals, and occurring in various other microhabitats; (2) larger, multicellular and macroscopic algae in a variety of forms but especially as the large, leathery seaweeds of rocky outcrops and the finer, more filamentous species growing on the surfaces of coarser seaweeds or on rock; and (3) the sole examples of marine tracheophyte communities, the sea-grasses, salt-marsh herbs, and mangrove-swamp shrubs and trees (Fig. 1.19). The mat-forming or interstitial protists (and moneran blue-green algae), the fine, epiphytic seaweeds and the sporeling stages of the coarse wracks and kelps provide a direct source of food for many herbivores, especially gastropod molluscs, whilst the larger seaweeds and the tracheophytes on their decay form abundant detritus ultimately consumed by suspension and deposit feeders.

The constraints on size of individual primary producers do not apply in the littoral fringe of the seas and, for example, giant kelps such as *Macrocystis* can achieve total lengths in excess of 50 m, can grow at a rate of 25 cm per day, and can lead to extensive areas of coast being visibly dominated by photosynthetic organisms. The kelps and other large seaweeds although growing from the substratum are not rooted in it; their holdfast system serves only an anchorage, not a nutritive function. They, and the tracheophytes, are consumed by relatively few species, most notably by some sea-urchins and opisthobranch sea-hares. Thorson (1971) has given a graphic description of the consumption of the giant kelp *Nereocystis* by a Californian sea-hare, *Aplysia*: 'If one gives them a stem ... the thickness of a broomstick one can hear their jaws crunch away at the crisp algal tissue, rather like a child chewing a raw carrot; half a metre of stem is probably regarded as a reasonable meal every 24 hours'. Kelps also release copious dissolved organic substances.

The pelagic and benthic regions of the sea are most closely and most complexly interlinked in the littoral zone, and here all the rich plankton of neritic seas can be filtered from the water by benthic suspension feeders. It is therefore in the littoral zone that the benthos reaches its apogee in systems such as coral reefs, sheltered estuaries and backwaters, and rocky intertidal regions. Suspension feeders dominate. They display two features rarely seen in terrestrial animals. First, they are sedentary or completely

(a)

(b)

(c)

Fig. 1.19 (a) Virgin *Rhizophora* mangrove-forest, Obi Latu, Indonesia (World Wildlife Fund, copyright reserved). (b) Sea-grass meadow (*Enhalus*), Morotai, Indonesia (photograph and copyright N.V.C. Polunin). (c) Salt-marsh, Norfolk, Britain (photograph and copyright R.S.K. Barnes).

35

sessile. In order to find food, there must be relative motion between an animal and its food-bearing environment. For suspension feeders the food is contained within the water column and this may move, through tide or current, past the stationary animal, or if natural water movement will not suffice the suspension feeder can set up its own current by the beating of cilia, setae or other appendages. By living in a tube within the sediment and creating a current of water through that tube, an animal may derive the benefits of an infaunal existence whilst retaining respiratory and feeding access to water. Sessile suspension feeders are amongst the most inanimate of animals, at least as adults: the skeleton in animals is a structure facilitating locomotion, but in sponges it serves the converse function—it prevents the sponge from changing shape whilst the internal water current moves relative to the cells.

Secondly, suspension feeders or more generally small-particle feeders are often colonial in the sense that different individual modular units, or polyps, together form a much larger group-structure of characteristic shape. The colony is produced asexually, the founding polyp budding off new polyps which themselves bud, and so on, individual polyps often retaining tissue contacts with those nearest them. This is an analagous process to the creation of a metazoan body by the repeated asexual division of a founding zygote. And, as in the metazoan body, division of labour amongst the asexual division products may occur. This reaches its height in bryozoan and hydrozoan colonies (especially the pelagic siphonophores), in which feeding, reproduction, defense and, in the siphonophores, movement and flotation are carried out by different polyps or aggregations of polyps. Such colonies have a considerable claim to be considered as individual organisms in which there are two heirarchies of organizational levels: polyps formed of differentiated cells; and the colony/individual formed of a differentiated series of polyps.

1.4 Latitudinal gradients and global patterns

The third of the three great gradients affecting marine organisms was omitted from consideration on pp. 4–16 and can conveniently be discussed now. Solar energy is not distributed evenly across the surface of the globe. Regions within the tropics experience per day 12 hours of daylight and 12 hours of darkness, the light being intense and the heat input being large. In higher, temperate latitudes, summer daylight exceeds a duration of 12 hours per day and winter daylight periods are less than 12 hours, the light received is of lower intensity and the overall heat input is lower. Finally, in polar regions, a six month period of continuous, low-level daylight alternates with six months of darkness, temperatures being low throughout the year. This global pattern of light and heat receipt is of profound importance to the photosynthetic potential of the photic zone, both via its direct affects on photosynthesis and via the establishment of thermoclines.

If the light energy received by surface waters is sufficient for the phytoplankton to fix more carbon per day in photosynthesis than they require to respire in that period, then the heat received at the same time will be of the order of magnitude to generate a thermocline. Thus, in open tropical waters, light intensities will permit phytoplanktonic production to last throughout the year but a deep and permanent thermocline will also persist throughout all months. Light energy input may be favourable but nutrient availability will be unfavourable, and a low continuous level of production will result, totalling only some 50–100 mg C fixed m^{-2} of surface daily.

Moving further from the equator into temperate zones, seasonality becomes evident. During the winter months, the light intensity is so low that the photic zone is extremely shallow and well above the depth to which wind-induced mixing occurs. Therefore, although in the absence of a marked thermocline nutrients are plentiful, there is little or no production during the winter. At the start of spring, nutrients are still abundant in surface waters and as soon as the relationship between the depths of mixing and the photic zone becomes favourable, an outburst of production can take place. As this is happening, however, so is a seasonal thermocline being established and by sometime in summer the trapped stocks of nutrients will have been diminished and/or grazing pressure built up with the result that the outburst is ended. At the end of the summer the seasonal thermocline breaks up allowing the ingress of more nutrients, and if the light/mixing regimes remain favourable for a time before the onset of winter conditions, a second autumnal outburst of production may take place. This pattern of phytoplanktonic production need not, however, be reflected by a similar pattern of algal biomass (see pp. 55–6).

In even higher latitudes, nutrients are always abundant in surface waters but only during five months of the year is sufficient light available, and for the first and last of these months the relative depth of mixing is unfavourably large. Therefore the period of photosynthetic production is limited to only some three months of the year, but when for a brief time light, nutrients and

surface mixing are all favourable an outburst of massive proportions can occur. The magnitude of the summer period of production in high latitudes is sufficiently large to yield a higher
daily average productivity (150–250 mg C m^{-2} daily) than in
tropical open waters, in spite of the occurrence of unproductive
months.

This analysis has been based solely on the distribution of solar
energy to the open waters of the ocean, but the proximity of
coasts and of areas of upwelling are also relevant and they are not
distributed evenly across the globe either. Coastal waters, for example, occupy an increasing percentage of the sea as one pro-
gresses northwards from the equator, and we have already seen
that nutrients are more plentiful over continental shelves. These
enable neritic waters to support productivities in the range 125–
750 mg C m^{-2} daily, with an average value of the order of
450 mg C m^{-2} daily, or some three times the productivity of the
adjacent areas of oceanic water. Tropical neritic waters receiving
more light are, of course, more productive than those in higher
latitudes (all things being equal).

The second factor impinging is upwelling, and on pp. 11–12 it
was indicated that upwelling is particularly marked where water
is driven away from a coastline and where surface water masses
diverge. The force moving water away from a coast is wind, and

Fig. 1.20 The main areas of upwelling in the world ocean. (After Wright
1977–78).

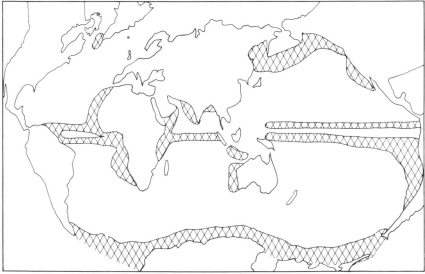

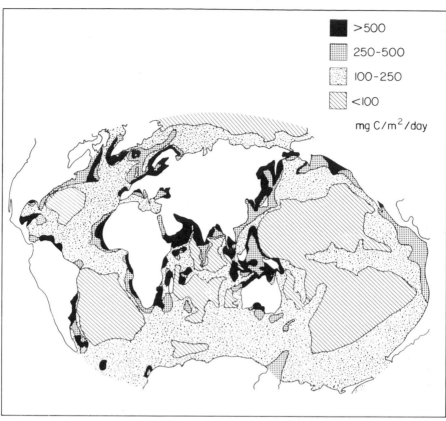

Fig. 1.21 The distribution of phytoplanktonic primary productivity in the world ocean. (After Koblentz-Mishke *et al.* 1970.)

the globe is subject to two broad belts of persistent winds which cause upwelling: the north-east trades in the Northern Hemisphere and the south-east trades in the Southern (Fig. 1.6). These have the effect of driving water away from the western coasts of continents in their path (and onto eastern coasts). Water therefore upwells along the western coast of north-west and south-west Africa, the western coast of the USA and of central South America, and the coast of west Australia; whilst the major zone of upwelling as a result of divergence is along the equator (see pp. 11–12). In addition, we saw (p. 14) that upwelling is associated with the generation of density-driven current systems in high latitudes, so that water upwells around Antarctica and, to a lesser extent, in the Arctic. Thus the distribution of upwelling in the world ocean is as shown in Fig. 1.20. The abundance of nutrients in these systems is sufficient to permit an average daily productivity in the range 500–1250 mg C m^{-2}, with an overall average of some 625 mg C m^{-2} daily.

Combining the latitudinal gradient, coastal effects and zones of upwelling, a global distribution of pelagic primary production of distinctive pattern results (Fig. 1.21). A map of the world is on too large a scale to show the influence of the littoral zone on global productivity, but for the purposes of comparison with the figures given above, the productivity of benthic algae and plants lies in the range 500–25 000 mg C m^{-2} daily (more usually within 2500–5500 mg C m^{-2} daily): benthic production per unit surface area exceeds pelagic production by a factor of ten. Although they can only inhabit less than 0.5% of the surface area of the oceans, by virtue of their large size coastal macrophytes also account for two-thirds of the total biomass of marine photosynthetic organisms. The existence of productive coral atolls in the middle of the barren tropical oceans is a paradox which will receive special consideration in Chapter 6.

It is not only overall productivity that varies with latitude, so do the nature of the organisms achieving this production and the characteristics of the assemblages of species supported. Amongst the pelagic algae, for example, diatoms make up a significant fraction of the total only in high latitudes, along the equator and off western continental margins, i.e. in zones of abundant nutrients. Conversely, dinoflagellates, coccolithophores, blue-green algae and several nanoplanktonic groups are most important in tropical, open-ocean waters. Other distribution patterns can be correlated directly with nutrient levels—that of the larger benthic kelps centres on coastal regions of upwelling—but not all latitudinal patterns are so easily explained. Whole habitat types may show a tropical versus high latitude division: coral reefs and mangrove-swamps are tropical whereas salt-marshes replace mangrove-swamps in temperate and boreal zones. The same phenomenon is seen in individual systematic groups. The ocypodid and grapsid crabs which dominate many intertidal habitats in the tropics are absent from high latitudes, for example, even though apparently comparable habitat-types are still available. Attempts have been made to correlate such distribution patterns with a single environmental variable; temperature in respect of reef corals and rainfall for the mangroves, but the detailed patterns still remain largely unexplained.

Although not an explanation in itself, Thorson (1957) showed many years ago that if one compares similar coastal regions in different latitudes, the numbers of infaunal species present per unit area is almost constant, but the species richness of the epifauna increases with decrease in latitude. The observed pattern of species diversity in the coastal macrobenthos (Fig. 1.22) is largely if not entirely attributable to the increase in the epifaunal component alone. Coral reefs and mangrove-swamps are two of the spatially complex tropical systems which (other things being equal)

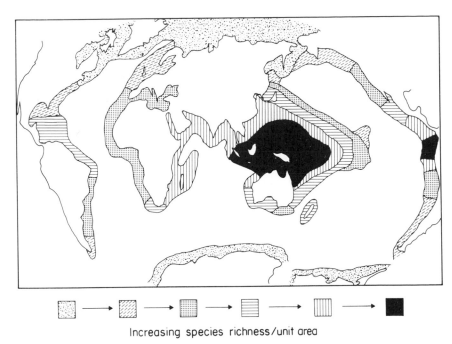

Increasing species richness/unit area

Fig. 1.22 The relative species-richness of different shallow-water regions of the world ocean. (After Valentine & Moores 1974.)

have permitted large numbers of epifaunal species to adapt to small microhabitats and to coexist, but this does not account for the absence of reefs and mangrove-swamps from more temperate conditions. Both are, however, mainly epifaunal phenomena and it is probably no accident that the infauna which appear insensitive to these latitudinal effects are the component least exposed to the extremes and fluctuations of the shelf/littoral environment. The infauna are cushioned by living within the sediments and the less labile interstitial habitat. This story will be pursued further in Chapter 3.

This chapter has set out to introduce the sea and its organisms, and to provide a very broad, overall picture of marine primary production. In the following seven chapters we will concentrate on the more detailed patterns and processes characterizing the major subdivisions of the marine environment outlined above. The world ocean, however, forms one large ecosystem in that all its component habitats interlink to form a three compartment system—pelagic : benthic : fringing—typical of all aquatic environments (Barnes & Mann 1980). Chapter 11 will examine the nature of these interactions and their magnitudes.

2 The Planktonic System of Surface Waters

It was emphasized repeatedly in Chapter 1 that several processes are confined to the surface waters of the seas, although the depth down to which each penetrates varies with the nature of the process. In this chapter we will consider the water column above the arbitrary depth of 1000 m so as to include all potential surface effects; deeper waters will be covered in Chapter 7.

2.1 The nature of pelagic photosynthesis

Light energy in the sea is used by organisms to synthesize complex organic molecules in three rather different processes, of which two are confined to special habitat types. Where simple, dissolved organic compounds (e.g. glucose) are already abundant in the water—as a result of previous photosynthetic activity—these can be converted into more complex substances by the process of 'photoassimilation'. In this, glucose, acetate, etc. replace carbon dioxide as the carbon source, but otherwise the process is similar to the photosynthetic pathways described below. True photosynthesis involves the introduction of hydrogen into the carbon dioxide molecule to form compounds with the empirical formula of $n(CH_2O)$, i.e. carbon dioxide is reduced or hydrogenated. In regions containing little oxygen, several bacteria and in some circumstances various protists, use compounds such as hydrogen sulphide as the hydrogen donor and photosynthesize according to the equation:

$$CO_2 + 2H_2X \longrightarrow CH_2O + 2X + H_2O \tag{2.1}$$

where H_2X is any reduced molecule other than water. The most frequent process in the oceans as a whole, however, is the photosynthetic fixation of carbon by the phytoplanktonic protists using water as the hydrogen donor, in which the X of equation 2.1 is replaced by O:

$$CO_2 + 2H_2O \longrightarrow CH_2O + O_2 + H_2O \tag{2.2}$$

This process liberates oxygen, which is not the case in bacterial photosynthesis.

Algal photosynthesis is therefore basically the same process as that in terrestrial plants and Whittingham (1974) or other botanical texts may be consulted for the biochemical details (see also Parsons *et al.* 1977). (It should be noted, however, that equations 2.1 and 2.2 are empirical simplifications of what in reality are a complex series of biochemical reactions, only one of which, that dissociating the hydrogen donor to release hydrogen ions and free electrons, requires light energy.) The light of 400–720 nm wavelength which powers the essential first stage of photosynthesis is absorbed by various photosynthetic pigments which are responsible for the characteristic colours of the phytoplankton. In the dominant protists (though not in the photosynthetic bacteria), the most important of these pigments is chlorophyll a which exhibits peak absorption of light of 670–695 nm wavelength. However, the phytoplankton possess a host of other accessory pigments (Table 2.1) which absorb light of shorter wavelengths, in part using it directly to dissociate water molecules and in part passing the light energy to chlorophyll a.

The wide range of light-absorbing pigments present in planktonic (and benthic) algae enables a broad band of wavelengths to

Table 2.1 Photosynthetic pigments of marine phytoplankton.

	Cyanophyta	Chlorophyta	Xanthophyta	Chrysophyta	Bacillariophyta	Cryptophyta	Dinophyta	Euglenophyta	Prasinophyta	Haptophyta
Chlorophyll a	+	+	+	+	+	+	+	+	+	+
Chlorophyll b		+						+	+	
Chlorophyll c			?	+	+	+	+			+
α carotene		+				+			+	
β carotene	+	+	+	+	+		+	+	+	+
γ carotene		+							+	
ε carotene					+	+				
Fucoxanthin			+	+	+		+			+
Neofucoxanthin				+	+					+
Diadinoxanthin			+	+	+		+			+
Diatoxanthin			+	+	+					+
Dinoxanthin							+			
Peridinin							+			
Neoperidinin							+			
Lutein	+	+	+						+	
Zeaxanthin		+							+	
Flavoxanthin		+								
Violaxanthin		+	+	+					+	
Neoxanthin		+	+					+		
Alloxanthin						+				
Monodoxanthin						+				
Crocoxanthin						+				
Siphonoxanthin									+	
Myxoxanthophyll	+									
Myxoxanthin	+									
Anthraxanthin	?									
Astaxanthin		+						+		
Oscilloxanthin	+									
Echinenone	+							+		
Phycocyanins	+						+			
Phycoerythrins	+						+			
Others		+	+				+	+	many	+

43

be used in photosynthesis, and it has also permitted different protist groups to specialize in the use of different wavelengths and therefore to trap light not absorbed by other species. In particular, phytoplankton species inhabiting relatively deep sections of the photic zone are adapted to the prevailing wavelengths by the possession of the most appropriate pigments. The photosynthetic abilities of a given species can be encapsulated in an 'action spectrum' which combines information on the wavelengths which can be absorbed and on the amount of carbon fixed by unit quantity of the light captured by each pigment (the efficiency of energy transfer to photosynthesis varies with the nature and functional role of the various pigments). Representative action spectra are shown in Fig. 2.1.

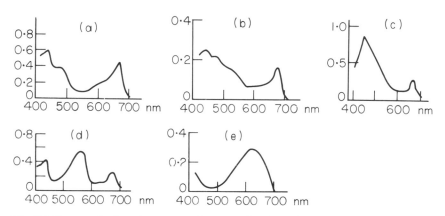

Fig. 2.1 Photosynthetic action spectra of some marine phytoplankton: (a) *Chlorella*; (b) *Gonyaulax*; (c) *Coccolithus*; (d) *Phormidium*; and (e) *Cyanidium*. (After Parsons *et al.* 1977.)

This photosynthetic fixation is responsible for the primary generation of organic compounds in the sea. Carbohydrates, fats and proteins are all synthesized and the total quantity of carbon or energy fixed forms the 'gross primary production'. The latter suffers three immediate fates. Some of the gross production will be broken down by the respiratory metabolism of the photosynthetic organism itself. Some will be incorporated into its tissues and fluids, and thereby constitute growth. Some will leak from the alga into the surrounding water and there join a pool of dissolved organic substances. 'Net primary production' is the term usually given to that proportion incorporated into the protist's body and therefore available to herbivores (mainly because of the importance which historically has been attached to the primary producer → herbivore food chain). However, net production should also include the loss of organic substances to the water as they are also utilized by consuming species, especially by bacteria but also by some zooplankton. All phytoplankton are leaky—a phenomenon utilized, and magnified, in various symbiotic relationships (see Chapter 6)—and in nutrient-poor waters the loss may be equivalent to more than 40% of the gross primary production. Release of glycollic acid may result in concentrations of this substance of 0.1 mg l^{-1} in sea-water.

2.2 Factors limiting primary production

Chapter 1 identified four factors as impinging on the magnitude of primary production achieved in the sea—light, mixing, nutrient availability and grazing—and we will now investigate these in some detail.

2.2.1 Light

The relationship of photosynthetic production to the rate of supply of light energy ('irradiance' or 'intensity' as measured, for example, in Watts per unit area or Langleys per unit time) takes the general form shown in Fig. 2.2. With increase in light intensity from zero there is first a linear phase of increasing photosynthesis, then a plateau is attained, and finally photosynthesis decreases at high intensities. During the linear phase, light is limiting and production is proportional to light intensity; the photosynthetic mechanism is saturated along the plateau; and harmful effects of too intense a supply of radiation set in at the start of the decline and increase thereafter. These harmful effects are both real and apparent. The real ones are due to the destructive action of ultraviolet radiation and to an overflow of light energy into oxidation processes (photo-oxidation); in bright sunlight these account for most of the decline. To some extent, however, the decrease in photosynthesis may only be apparent in

Fig. 2.2 The relationship between light intensity and photosynthetic production in planktonic diatoms (mean of five species). (After Ryther 1956.)

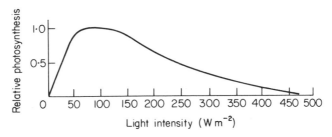

that: (a) a light-stimulated increase in respiration, 'photorespira-tion', may occur, resulting in an increased metabolism of fixed compounds; and (b) if high light intensity is coupled with short-age of the nutrients also required in photosynthesis (see pp. 49 *et seq.*), the rate of leakage of fixed products from the alga is in-creased, sometimes equalling 90% of the gross production.

If we apply this production–light-intensity relationship to the surface waters of the sea (through which light decreases exponen-tially with depth), the pattern displayed in Fig. 2.3 would be ex-pected. Near the surface, excess light will result in photo-inhibition whilst at depth, light intensity will quickly decline to limiting values. Average values for the critical points on this dis-tribution would lie in the general area of 170 W m^{-2} for the onset of photoinhibition and 120 W m^{-2} for saturation. Different species and different algal groups, however, display differing characteristic values; in diatoms, for example, both saturation and inhibition occur at lower light intensities than in dinoflagel-lates.

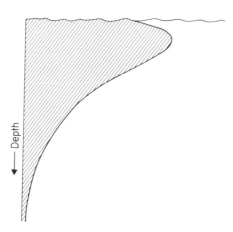

Fig. 2.3 The relationship between depth and photosynthetic production in the surface waters of the ocean.

A distribution of photosynthesis with depth in the sea similar to that of Fig. 2.3 is indeed found in nature, with minor modifi-cations resulting from the distribution in the water column of the photosynthetic organisms (Fig. 2.3 assumes a uniform distribution). On relatively dull days, of course, there will be no photoinhibition; and light penetration may be curtailed by self-shading if dense congregations of algae are present near the sur-face. Nevertheless, constancy in the general shape of this curve (especially when plotted as percentage of the maximum production in the water column against a measure of light penetration) has permitted the use of formulae in the calculation of the total gross photosynthetic production (ΣP) achieved beneath each square metre of sea surface. One such is:

$$\Sigma P = nP_{max}d \qquad\qquad (2.3)$$

where nP_{max} is the rate of maximum photosynthesis (photosynthetic rate per unit population at saturation multiplied by phytoplankton density) and d is the depth at which the intensity of the most deeply penetrating wavelength is reduced to 10% of its surface value. It is therefore customary to express production per unit area of surface, not per unit volume.

In contrast to photosynthesis, respiration rate does not vary with depth (except in the uppermost layers where it may be enhanced by photorespiration) and it is usually considered to dissipate some 10% of the gross primary production achieved at saturation. Hence if allowance for respiration is made on the curve relating photosynthesis to depth (Fig. 2.4), a point will occur at which photosynthetic fixation and respiratory dissipation of

Fig. 2.4 The relationship between depth, phytoplanktonic photosynthesis and phytoplanktonic respiration in the surface waters of the ocean, showing the compensation and critical depths.

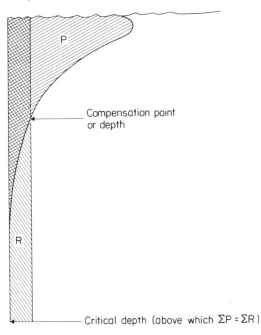

carbon are in balance. This is known as the 'compensation point' or depth. Clearly it will vary in the water column with changes in light intensity, so it is usual to express the compensation point as the depth at which the carbon fixed (or oxygen produced) in photosynthesis over a period of 24 h is equal to the carbon dissipated (or oxygen consumed) in respiration over that same period. The depth, of course, may still vary seasonally. Above the compensation point, phytoplankton may grow and multiply; below it, they must either subsist on accumulated reserves, form inactive resting bodies, or starve. Although expressed as a depth, the real variable involved is light intensity and hence the corresponding level of irradiance is known as the compensation light intensity. In practice, rules of thumb are used to establish the compensation depth, rather than precise measurements of photosynthesis and respiration. One such rule places it at the depth to which 1% of surface light penetrates and the same depth then separates the photic and aphotic zones (see p. 4).

2.2.2 Turbulence

At any one time, the compensation depth is a real balance point in the metabolic physiology of a phytoplanktonic alga. A second point which can be derived from graphs of respiration and photosynthesis against depth is a purely notional one but it is nevertheless vital in the relationship between light penetration and the zone of surface mixing. A depth will occur in any water mass above which the gross primary production of carbon in that water column is equal to the total respiratory release in the same column (Fig. 2.4); i.e. the excess of production over respiration above the compensation point is balanced by the excess of respiration over production below it. This second point is known as the 'critical depth' and in terms of irradiance it is the depth above which the average light intensity is equal to the compensation light intensity. It can be calculated from a knowledge of the compensation light intensity, the light intensity at the surface and the extent of light penetration.

On average, a photosynthetic organism can circulate through the water column above the critical depth whilst still being in photosynthetic—respiratory balance over 24 hours. Wind-induced turbulence may extend down to depths of 200 m yet the photic zone may be much more shallow (p. 4). It follows from the above that net photosynthetic production will only be possible if the depth to which mixing takes place is higher in the water column than the critical depth (Fig. 2.5). If mixing extends

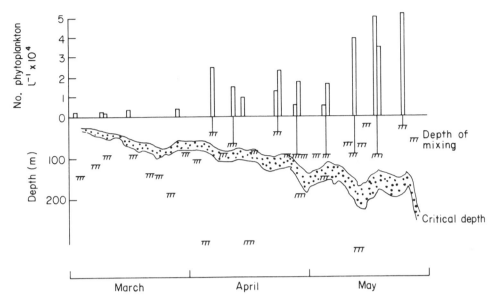

Fig. 2.5 The relationship between phytoplanktonic abundance and: (a) the depth to which the surface layers are mixed; and (b) the critical depth. (After Sverdrup 1953.) Note that phytoplankton are only abundant when the depth of mixing is less than the critical depth.

below the critical depth, as it frequently does during the winter in relatively high latitudes, the average light intensity experienced by a photosynthetic organism will be less than the compensation light intensity and production will be negative until such time as circumstances change; for example when the depth of mixing decreases with abatement of wind velocity, or when light intensity increases on the approach of spring or summer.

2.2.3 *Nutrients*

In the discussion in section 2.2.1, photosynthesis was described in terms of the production of $(CH_2O)_n$ from CO_2 and H_2O (equation 2.2), but other elements are required by even the most completely autotrophic of organisms in that they comprise parts of proteins, enzymes, energy-stores, energy-carriers and other molecules (Stewart 1974). Hence in order to grow and metabolize, both energy and the necessary building blocks and energy-carriers must be available. Most of these material requirements are available to excess in sea-water, but concentrations of nitrogen and phosphorus in particular can at times and in certain areas be very low.

In addition, a number of photosynthetic organisms are not completely autotrophic. Dinoflagellates and various other algae (particularly members of the Chrysophyta, Cryptophyta and Haptophyta), and some species in all algal groups, require external sources of certain organic compounds, especially vitamins, since they are unable to synthesize them themselves. It has been known for many years that in order to culture these 'auxotrophs', addition of such materials as humic acids must be made to the culture medium.

Analysis of the extent to which nutrient shortage limits marine photosynthesis might appear, in theory, to be a relatively simple matter. In fact it is fraught with conceptual and investigative difficulty and before presenting the evidence it will be helpful to review the nature of the problems. One major difficulty centres on the observation that different species have differing abilities with which they can take up environmental nutrients, and indeed have differing requirements for these nutrients. Uptake of nutrients is an active process that can operate against a concentration gradient but which is nevertheless dependent on the external concentration. Rather as with light absorption, at low external concentrations of a nutrient, uptake is dependent on concentration, but at a certain (higher) level the uptake mechanisms saturate and a plateau is attained (Fig. 2.6). The rate of uptake (U) can therefore be expressed as

$$U = \frac{U_{max}C}{C + K_c}. \tag{2.4}$$

where U_{max} is the rate at saturation, C is the concentration of the nutrient in question, and K_c is a constant equal to the nutrient concentration at which U equals $0.5U_{max}$. The half-saturation constant, K_c, is an indication of the ability of a given photosynthetic organism to take up that particular nutrient from a

Fig. 2.6 The general relationship between the external nutrient concentration and the rate of uptake of that nutrient by algae.

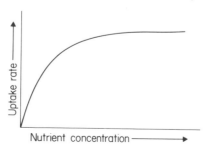

Table 2.2 Half-saturation constants for nitrate and ammonium uptake. (From data in Eppley *et al.* 1969 and MacIsaac & Dugdale 1969.)

Phytoplankton from	K_c nitrate (μg-at. 1^{-1})	K_c ammonia (μg-at. 1^{-1})
Nutrient-poor areas of the tropical Pacific	0.01–0.21	0.10–0.62
Nutrient-rich areas of the tropical Pacific	0.98	
Nutrient-rich areas of the polar Pacific	4.21	1.30
Oceanic in general	0.1–0.7	0.1–0.4
Neritic (diatoms)	0.4–5.1	0.5–9.3
Neritic and littoral (flagellates)	0.1–10.3	0.1–5.7

range of external concentrations. (In fact, K_c is not truly constant; it varies with temperature and with the internal nutrient concentration of the alga.)

The point at issue, however, is that different members of the phytoplankton have widely varying values of K_c (Table 2.2), such that some species can take up nutrients only from high external concentrations whilst others can do so from progressively lower concentrations. Thus whilst it is easy to see how falling nutrient concentrations (resulting, for example, from uptake from a finite pool trapped above a thermocline (see p. 53)) could limit the photosynthetic production of a *given* phytoplanktonic *species* with a high nutrient demand, it is less easy to see how this could limit phytoplanktonic production *in total*. What happens in such circumstances is that local selection pressures now favour species which can take up nutrients from lower external concentrations, and as concentrations drop still further so yet different species will be at an advantage, and so on. This process is reflected by the fact that the phytoplankton species dominating a given water mass at a given time have K_c values appropriate to the ambient nutrient concentrations (Fogg 1980). It is also evidenced by seasonal successions of different phytoplankton (pp. 58–9). Therefore it is only by demonstrating that the phytoplankton characterizing waters of low nutrient status are, and must be, inherently less productive than those of more nutrient-rich areas, that a nutrient limitation on phytoplanktonic production in total can be established. This does in fact appear to be the case. It must be emphasized, however, that demonstrations of nutrient limitation (whether mineral or organic) of individual species are not strictly relevant to nutrient limitation of production in general.

A second factor complicating any straightforward relationship between nutrient concentration and productivity is that growth and multiplication rates are related to the concentration of nutrients within the primary producer not to those in the surrounding water. Phytoplankton can store nutrients taken up at times of relative plenty and use them for subsequent production even in the absence of external supplies. From two to more than five further generations may be fuelled from stored sources.

Such complications notwithstanding, there is abundant evidence of a general relationship between external nutrient supplies

Fig. 2.7 The general correspondence between: (a) the distribution of nutrients, in this case of phosphates at 100 m depth (after Reid 1962); and (b) of phytoplanktonic productivity in the same ocean (the Pacific). (From data in Koblentz-Mishte *et al.* 1970.

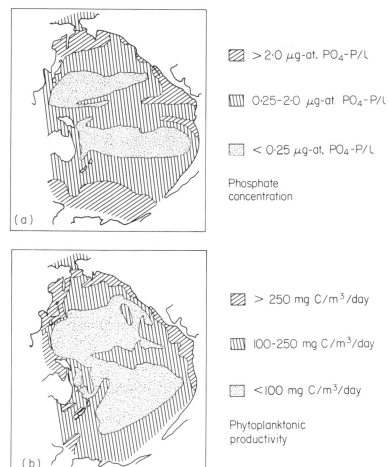

$> 2 \cdot 0 \ \mu g\text{-at. } PO_4\text{-P}/l$

$0 \cdot 25 - 2 \cdot 0 \ \mu g\text{-at. } PO_4\text{-P}/l$

$< 0 \cdot 25 \ \mu g\text{-at. } PO_4\text{-P}/l$

Phosphate concentration

$> 250 \ mg \ C/m^3/day$

$100 - 250 \ mg \ C/m^3/day$

$< 100 \ mg \ C/m^3/day$

Phytoplanktonic productivity

and phytoplankton biomass and productivity, largely in the form
of correlations between the two in both space and time. We saw
in Chapter 1 that the global pattern of productivity correlated
well with nutrient levels, being high in the nutrient-rich shelf
waters, in areas of upwelling and during the absence of a thermocline (solar radiation permitting), and low in the nutrient-poor,
tropical oceanic regions. A more detailed example of such a spatial correlation is shown in Fig. 2.7. Similarly, in some areas with
seasonal climates nutrients are known to be depleted during the
presence of the warm-season thermocline and production also decreases; both nutrient levels and productivity increase again
when the thermocline disappears (Fig. 2.8). Finally, it is known
that when sea-water is enriched artifically with nutrients—as
when sewage is discharged into semi-enclosed bays—primary
production is stimulated, suggesting that nutrients were limiting
hitherto.

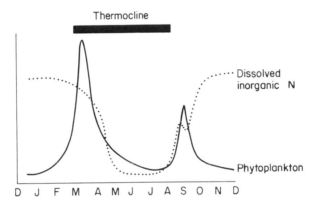

Fig. 2.8 An idealized relationship between phytoplanktonic biomass in the North
Atlantic Ocean and: (1) the external concentration of nitrogen; and (2) the
duration of the seasonal thermocline.

Early work concentrated on the potential limiting role of phosphorus since inorganic phosphates are less abundant in sea-water
than ammonia and nitrates, the forms in which most phytoplankton require their nitrogen—phosphorus and nitrogen atoms—are
present in the ratio of 1 : 15. In part the bias in favour of phosphorus was also a reflection of the availability and sensitivity of
techniques for the routine determination of the two elements.
However, with the discovery that several algae possess alkaline
phosphotases and could obtain their phosphorus from the more
abundant stocks of dissolved organic phosphates, attention
switched to nitrogen and this element is now generally accepted

as that associated with most cases of nutrient limitation (see also p. 83). Indeed it may well be significant in this context that some of the most successful algae in areas or at times of extreme nutrient poverty are the moneran blue-greens, some of which can obtain their nitrogen from dissolved nitrogen gas, of which there is no shortage in the sea. (Nitrogen fixation by blue-green algae is a feature of great ecological importance in other nutrient-poor tropical waters as well, for example on coral reefs (see p. 191)).

Phytoplankton are small and can effectively take up nutrients only from the thin layer of water surrounding them. Without relative motion, this thin layer would soon be impoverished regardless of the overall concentrations of nutrients in the larger water mass in which they are suspended. Motile flagellates can change water masses on a microscale and dinoflagellates can even maintain a movement of water over their surface by the beating of the transverse flagellum. This may be the main selective advantage of the possession of flagella. The immotile diatoms, however, have no means of maintaining conditions favourable to nutrient uptake over their surface. The tendency to sink as a result of gravitational forces (p. 23), rather than being completely disadvantageous appears necessary to the maintenance of suitable nutrient-uptake gradients in these species.

How then can we explain the mode of action of nutrient shortage in limiting overall production? As suggested above, the solution to the problem is most likely to be found in the nature of the phytoplankton species which are at a selective advantage under different nutrient regimes, and we can best illustrate this by examining the two extreme cases. In areas or at times of nutrient abundance, there will be no premium on efficiency of nutrient uptake and species with high half-saturation constants will not be at any selective disadvantage. Except in areas of permanent nutrient abundance in the tropics, however, conditions conducive to photosynthesis will not persist: concentrations of nutrients may be reduced by utilization or the input of light energy will decline. Hence the period of nutrient abundance is a temporary bonanza which, as in all similar circumstances, is best exploited by an r-strategy (i.e. one that maximizes the potential rate of population increase (see p. 217)). Under these conditions, it is often light which is limiting as a result of self-shading. Therefore there will be selective advantages associated with a 'bloom life-style': species will be favoured which multiply very rapidly in order to pre-empt the light and utilize the nutrients whilst they are available; and anti-herbivore defences will not be required as herbivore build-up will either be slow in relation to the rate of increase of the blooming phytoplankton (p. 56) or may even be non-existent (p. 65). In other words, species which devote a large proportion of their available energy to rapid growth and multiplication and 'make hay while the sun shines' (i.e. are highly productive) will not be at a selective disadvantage; indeed, they will probably be at an advantage in that they can out-compete more slowly increasing species for the available light.

Turning to the other end of the spectrum, productive, opportunistic species will be unable to survive in a system in which there is a premium on efficiency or on quality rather than quantity. High half-saturation constants may enable large quantities of nutrients to be taken up rapidly to fuel a high nutrient demand economy, but such an economy cannot be maintained in areas of resource shortage. In the nutrient-poor surface waters of the tropics, light may be abundant but competition for nutrients will be intense, as will competition amongst consuming species for their food (p. 71). Here there will be a premium on the efficient utilization of resources and on avoidance of being consumed. Selection will, at this end of the spectrum, favour species which devote a smaller proportion of their energy to production: it is the species that can survive best on least which will be the victors of competitive battles. The result is an assemblage of light-demanding, K-strategists efficient at removing nutrients from low concentrations and using them with maximum metabolic efficiency (K being the symbol for 'carrying capacity' and indicating that when resources are scarce the ability to compete effectively is more advantageous than a high value of r).

2.2.4   Grazing

The final potential check on phytoplanktonic production is the consumption of algal biomass by herbivores. It is clear that herbivores can have a marked effect on phytoplankton biomass. If, using known figures for the multiplication rates of algae or estimates of the quantities of nutrients taken up (as evidenced by decreases in environmental nutrient concentrations), one calculates the quantity of algal biomass that should have been produced during unit time, one finds that the actual biomass present is very low in comparison—perhaps as little as 0.5% of that expected (Fig. 2.9). In relatively stable areas, herbivores may remove phytoplankton production as fast as it is formed. This is not always the case however. In high latitudes in which there is a summer bloom of phytoplankton, build-up of herbivore numbers may lag some six weeks behind that of their food (Fig. 2.10). The

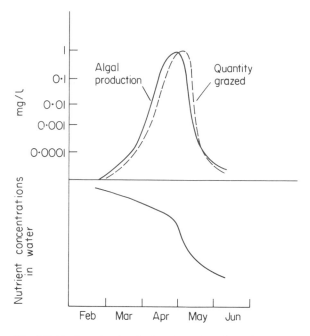

Fig. 2.9 Algal production during the spring phytoplankton bloom in the temperate North Atlantic Ocean in relation to the decline in the external nutrient levels caused, and the grazing pressure exerted by herbivorous zooplankton. (After Cushing 1959.)

herbivorous zooplankton of markedly seasonal areas often over-winter in resting stages and do not begin their annual breeding cycle until they have fed on the algae available after the bloom has started. The delay in increase of herbivore densities and the large difference between the potential growth rates of algae and herbivores ensure the production of large quantities of algal tissues which are not taken by consumers. No hard-and-fast latitudinal rule can be applied, however, because in some areas, the North Pacific for example, the dominant herbivores can 'predict' the onset of the bloom, breed in advance of its start using accumulated reserves, and make greater inroads into the biomass when it is produced.

The trophic inter-relations between algae and herbivores are by no means entirely one way. By consuming algal biomass and metabolizing their tissues, herbivores release nutrients from their organic binding and excrete them into the environment. Equivalently, by feeding on bacteria which have themselves absorbed the dissolved organics released by the phytoplankton, the microplankton release further stocks of nitrogen and phosphorus back

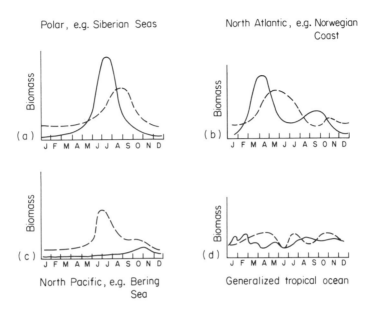

Polar, e.g. Siberian Seas

North Atlantic, e.g. Norwegian Coast

North Pacific, e.g. Bering Sea

Generalized tropical ocean

Fig. 2.10 Seasonal patterns in the biomass of phytoplankton (continuous line) and zooplankton (dashed line) in different latitudes and oceans. (a–c after Heinrich 1962; d after Parsons *et al.* 1977.)

into the water. These consumers thus regenerate incorporated nutrients and make them available to the phytoplankton again, thereby stimulating productivity. Indeed it is likely that in nutrient-poor areas, primary productivity might cease were the herbivores and bacterivores to cease regenerating nutrients *in situ*: they may stimulate as much productivity as they consume. The question therefore arises, Although herbivores may consume much, and in some cases the vast majority, of phytoplankton production, do they limit algal productivity as well as algal biomass? There is no general agreement on this question, nor indeed is there any reason to expect a single general answer. It is relatively easy to see that when a phytoplankton bloom has reached a limit set by decreasing light intensity or decreasing nutrient concentrations, the delays inherent in the herbivore response pattern can result in an eventual bias in favour of the herbivore populations and a crash of the algal populations. Here herbivores graze down a bloom which had already been stopped by other agencies. Whether, over large areas, herbivores can limit production which is not already at a ceiling as a result of unfavourable conditions of light, turbulence or nutrients is not so simple a matter. At some times, herbivore populations follow those of their prey in a typical Lotka–Volterra fashion; at other times, herbivore decline may actually preceed that of their food source. Further aspects of these interactions will be considered later in this chapter, but it is certainly safe to conclude here that herbivore pressure is one element in a kaleiodoscope of factors that influence marine photosynthetic populations: it is often very difficult to determine exactly which of the several factors is limiting production at any one time.

2.3   Distribution of phytoplankton in space and time

Global maps such as that of Fig. 1.21 are useful in summarizing broad latitudinal and regional trends, but they may give a most inaccurate impression of uniformity over large areas. The distribution of phytoplankton species and productivity on a small scale is in fact very patchy in both the temporal and spatial planes.

Mention of seasonal changes in phytoplankton has already been made and we can start by amplifying these patterns. The general form of seasonal successions of dominant species is shown in Fig. 2.11. Most of the available information relates to the microplanktonic species and the extent to which the nanoplankton mirror their patterns is unknown: those few studies which have been carried out indicate that they may not. The succession of microplanktonic species appears largely to be a result of changes in the quantity and quality of nutrients and of organic vitamins, etc. Auxotrophs, in particular, require previous conditioning of the water by other species before they can bloom. In the Sargasso Sea, for example, vitamins play an important role in determining the successional sequence. Diatom species requiring external sources of B_{12} dominate during the spring and greatly reduce its environmental concentration. They release vitamin B_1 into the water, however, and this is required by a coccolithophore, *Coccolithus*, which blooms when B_1 concentrations pass its threshold value. In turn, this alga releases B_{12} and so conditions the water for the succeeding diatoms. (B_{12} is also released by several planktonic bacteria.)

Not all external metabolites are beneficial or neutral to other organisms. Dinoflagellate species are notably associated with the release of toxins into the water (see p. 94) and several algae appear to release substances with bacteriocidal properties. The flagellate *Olisthodiscus* produces a tannin-like compound which stimulates the growth of a diatom when present in low concentrations but inhibits it in high concentrations.

Productivity also varies within single periods of daylight in all but polar latitudes. One might predict that levels of primary

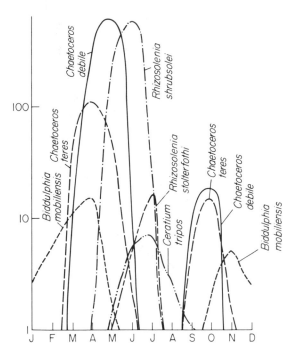

Fig. 2.11 Seasonal succession of dominant microplanktonic algae in the Irish Sea (approximate numbers in 1000s per m³ averaged over 14 years). (From data in Johnstone *et al.* 1924.)

production would vary with the diurnal change in light intensity and therefore be symmetrical about midday, but this is not the case. Photosynthetic production is usually greater before midday than after, by a factor of ten in tropical waters decreasing to zero near the poles. Many factors may interact to produce this pattern, but one important contributory cause is the tendency of the phytoplankton to divide at night. In the morning, the phytoplankton is dominated by young, growing, relatively productive individuals whilst by the afternoon the algae are older, slower growing and senescent.

In addition to this temporal variability, the phytoplankton is also distributed very variably in space: distinct patches occur, separated by intervening, relatively barren areas. The scale of the patchiness extends from the order of 100 km down to a few decimetres, big patches comprising aggregations of smaller patches. Three types of process appear to be responsible: physical, reproductive and feeding. The physical processes are those resulting in local turbulence, including the special case of Langmuir circulation. Weak to moderate winds blowing persistently across the

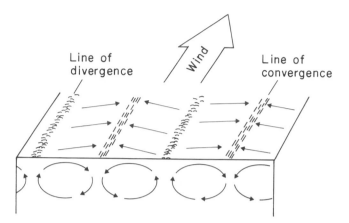

Fig. 2.12 The Langmuir circulation. Each vortex is *c.* 5 m in the shorter diameter and *c.* 15–30 m in the longer.

sea establish long, parallel, rotating cylinders of water, with adjacent cylinders rotating in opposite directions (Fig. 2.12). This Langmuir circulation therefore comprises alternate streaks of downwelling and upwelling which may extend for tens of kilometres and be separated by some tens of metres. Buoyant particles will aggregate in the downwellings and sinking particles in the upwellings; the aggregations representing concentrations of over one hundred times the background levels. Not only phytoplankton but zooplankton and pleuston become concentrated in these 'wind rows'. More generally, any purely local turbulence can raise nutrient concentrations over limited areas and there lead to increased phytoplankton production.

These physical processes are mainly responsible for large-scale patchiness, whilst the two biological mechanisms operate over smaller distances. On the smallest of scales, the division products of a single phytoplanktonic individual will tend to remain together and so magnify any original random unevenness of horizontal distribution. The other biological cause of patchiness is a product of the interaction between the grazing of herbivores and the biomass of their phytoplankton food (Fig. 2.13). Quite simply, in areas of greater than average herbivore density (in relation to the amount of food present) phytoplankton will be grazed down, whilst in areas of lower than average grazing pressure, phytoplankton biomass can increase. This will lead to patches dominated by zooplankton interspersed with ones dominated by algae. The system may well be dynamic, however, in that having grazed down an area the herbivores can through vertical migration (see pp. 24–5 and 67–70) move to another patch

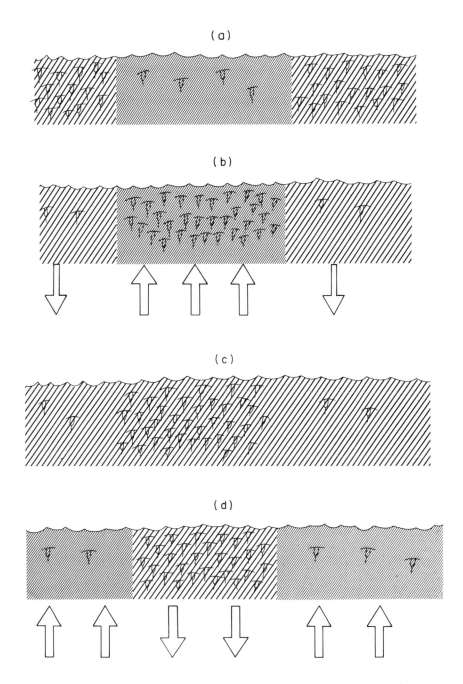

Fig. 2.13 A cyclic generation of patches dominated by zooplankton interspersed with patches dominated by phytoplankton. Patchiness is a consequence of the feeding behaviour and aggregation of the zooplanktonic herbivores. (After Bainbridge 1953.)

and proceed to graze that down too. Meanwhile, in the area vacated the phytoplankton can recover. In part, it is this cycle of grazing down and recovery, represented in space as a series of discrete patches along a phytoplankton-dominated to zooplankton-dominated continuum, which makes analysis of this, as of other such shifting aggregations, so difficult and so locally variable.

2.4   Zooplanktonic production

It has been estimated that 75% of pelagic primary production is dissipated in the uppermost 300 m and 95% in the top 1000 m, but 'in contrast with the frequency of primary production estimates in the sea, there are few good values for secondary production, and those that are available are rarely comparable with one another' (Tranter 1976). From the sparse data in the literature it does appear, however, that in general, secondary production reflects the distribution of primary production. Thus in the most coastal regions of neritic water and in some areas of upwelling, zooplanktonic production may average some 75 mg C m^{-2} daily, decreasing to 60 mg C m^{-2} daily in shelf waters generally, and to 12 mg C m^{-2} daily in tropical oceanic waters.

Unlike photosynthetic production, that of the zooplankton has rarely been measured directly and even then most measurements merely refer to single species. Figures such as those given above are obtained by multiplying values of primary production by some coefficient. That most frequently used is the 'ecological efficiency', defined as the amount of energy extracted from a given trophic level divided by that supplied to the trophic level. The ecological efficiency is itself the product of two other coefficients, the 'ecotrophic efficiency' (the proportion of the annual production of a trophic level taken by consumers) and the 'growth efficiency' (the annual weight increment of the consumers divided by the weight of food consumed). The production of a given trophic level (P) is therefore given by:

$$P = BE^n$$
$$\text{or } P = B (E_c G)^n \tag{2.5}$$

where B is the annual primary production fueling the system, E is the ecological efficiency, E_c the ecotrophic efficiency, G is the growth efficiency, and n is the number of trophic levels through which energy must pass to get to the trophic level of interest. One other efficiency may be mentioned at this point. The 'transfer efficiency', namely the ratio of the production of one trophic level to that of the next, is a reasonable estimate of the ecological efficiency (on the assumption that the energy extracted from a given trophic level is proportional to its production).

Unfortunately, several factors combine to render these methods of calculating secondary production open to question: no general agreement exists on the values to be assigned to E, and efficiencies can vary dramatically from area to area and from time to time; the concept of a trophic level is not one which can readily be applied to the marine ecosystem; and phytoplanktonic biomass is not the only food source available to the zooplankton. Therefore all estimates of zooplanktonic production (and of secondary production generally) are still not much more than guesses based on certain assumptions. However, because the assumptions, and possible criticisms of them, are if anything more instructive than are the estimates of production derived from them, we shall now investigate each in turn.

It is most convenient to start by considering the concept of a trophic level. This concept arose in terrestrial ecology to describe stages in the plant $\longrightarrow$ herbivore $\longrightarrow$ carnivore chain, and even in terrestrial ecology is now of dubious validity. This stems from two observations. First, food webs are almost everywhere extremely complex (Fig. 2.14) and, for example, what was identified as a sixth-trophic-level species in one marine study, because it fed in part on a fifth-level organism, also took animals in the fourth, third and second trophic levels. Secondly, in many ecosystems—and, it has been claimed, in all—some 90% of the energy flow does not derive directly from living photosynthetic organisms but passes along the detritus food chain in which the trophic level concept is even less applicable (since the primary detritus is itself an aggregate including organisms which could conceivably be placed in many trophic levels). A trophic level is therefore an abstraction bearing little relation to anything existing in the real world, and the marine system is no exception.

Values which different workers consider should be assigned to the efficiency of energy flow from consumed to consumer span a wide range, and the individual 'efficiencies' are themselves variable in nature for sound ecological reasons. This partly results from dissipation of energy by trophic interactions *within* a single notional trophic level, as a consequence of omnivorous diets or even cannibalism, but there are many other causes. Cushing (1971), for example, found that values of the transfer efficiency between phytoplankton and herbivores varied in a number of upwelling areas with the magnitude of the primary production. Efficiency was greatest ($c.24\%$) in regions of relative food scarcity

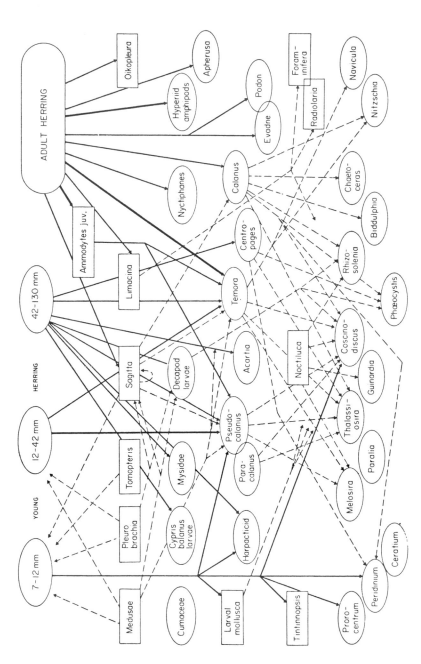

Fig. 2.14 The feeding relations of the herring, *Clupea harengus*, in the North Sea at different stages in its life cycle. (After Hardy 1924.)

Table 2.3 Gross growth efficiencies (energy from the food converted into new tissues in the consumer) of marine organisms, based on field and laboratory results. (After Pomeroy 1979.)

Species	% Efficiency
Bacteria	30–80
Oikopleura (pelagic tunicate)	4–50
Sagitta hispida (chaetognath)	25–50
Sagitta setosa (chaetognath)	9–18
Calanus hyperboreus, adult (copepod)	13–30
Calanus hyperboreus, stage IV–V copepodite larvae	15–50
Calanus finmarchicus, adult	14
Calanus finmarchicus, subadult	25–50
Calanus helgolandicus, nauplius—adult	18–72
Rhincalanus nasutus, nauplius—adult (copepod)	25–55
Euphausia pacifica (krill)	30
Palaemon adspersus (prawn)	1–10
Hediste diversicolor, adult (polychaete)	14–43
Mactra, 1–2 years old (bivalve)	55
Asterias (starfish)	55
Pleuronectes platessa, 0–1 years old (fish)	30
Pleuronectes platessa, over whole life	15

and least (*c.*3%) in areas of food abundance: a similar premium on efficiency to that noted on p. 55. Growth efficiency may vary within the herbivorous zooplankton from 7% to 50%, and ecotrophic efficiency between phytoplankton and herbivores may vary from only a few per cent (<5%) to over 90% (see Table 2.3).

This introduces one exception to the general relationship between phytoplanktonic and zooplanktonic productivity. In some very rich coastal areas, primary production may be of such large magnitude that, paradoxically, it is not consumed by pelagic organisms; it sediments out of the water column creating a large benthic oxygen demand and, occasionally, anoxia. Such examples are exactly comparable with the more familiar eutrophic ponds and lakes: zooplankton may actually be asphyxiated by the thick algal soup if they are not killed by lack of dissolved oxygen first. What little is certain in the field of trophic efficiencies is that the ecological efficiency of energy flow in the sea is generally higher than the 10% often quoted in terrestial ecology. At the phytoplankton–herbivore interface, it may lie in the vicinity of 20% (partly as a consequence of the greater digestibility of marine photosynthesizers), decreasing to between 10 and 15% further up the food web.

The third general assumption of equation 2.5 is that living phytoplankton are the main food source of pelagic consumers. There is no question that photosynthetic production is the base of the food web; the debate revolves around the quantities and proportion of living primary producers grazed. The alternatives are detritus and the bacteria subsisting on dissolved organics released by the phytoplankton. Bacteria are known to be consumed by heterotrophic protists which are themselves taken by small zooplanktonic animals. Similarly, faecal pellets, the gelatinous coverings of various members of the plankton, and dead algal cells are colonized by a variety of bacteria, protists and small metazoans (such as nematodes), and the whole detrital aggregates are eaten by many larger planktonic animals, some possibly subsisting entirely on this diet (see also pp. 78–9). Faecal pellets in particular are often very rich in food materials. Some phytoplankton appear capable of passing through copepod guts, without being digested, but, being encased in a heavy faecal pellet, they sink rapidly and join the detrital pool. A few even emerge alive and can wriggle free of the pellet and continue photosynthetic life. Although these alternative pelagic pathways to the grazing food chain are known, there is no concensus on their relative importance: estimates vary from more than 90% to an insignificantly small percentage of the total energy flow. Until such time as these and the other problems are resolved, there can be no certainty of the magnitude of secondary production nor of the detailed factors influencing it. All estimates must be treated with a modicum of suspicion.

A second important role performed by the zooplankton in the pelagic system has only been appreciated relatively recently. It was believed for many years that bacteria were the agents responsible for releasing from their binding those inorganic nutrients incorporated into organic tissues and thereby making them available again to the phytoplankton. Recently, however, it has been appreciated that bacteria themselves have a very high demand for phosphates and nitrates; indeed, a higher demand, weight for weight, than the phytoplankton because bacterial tissues contain more phosphorus and nitrogen atoms for every carbon atom than do those of the photosynthetic algae. Bacteria may be a sink for nutrients, not a source (although, nevertheless, bacteria are extremely important in changing the *form* in which elements such as nitrogen occur). It now appears more likely that it is the protist and animal consumers of phytoplankton and bacteria that are responsible for nutrient regeneration. Zooplankton excrete between 2 and 10% of their body nitrogen and 5–25% of their body phosphorus each day, with even higher values when food is relatively abundant. Recyling of these nutrients is therefore rapid; in temperate seas, for example, the turnover time of phosphate is only 1.5 days.

2.5 The spatial distribution of zooplankton

Zooplankton, like (and partly as a consequence of) the phytoplankton, are distributed patchily. Horizontal patchiness (Fig. 2.15) has already been discussed (pp. 60–2), but it is also relevant to the main subject of this section—a phenomenon which forms 'one of the most striking and characteristic aspects of the behaviour of marine zooplankton' (Raymont 1963). Most, though not all, members of the zooplankton undertake diurnal vertical migrations through the water column, of somewhat less than 400 m on average in the smaller species and of over 600 m (up to some 1000 m) in the larger. These vertical movements, which may involve sustained upward swimming speeds of 12–200 m h^{-1} dependent on size, and downward speeds some three times faster, may be undertaken twice each day.

Fig. 2.15 Horizontal patchiness in four species of zooplankton. (After Wiebe 1970.)

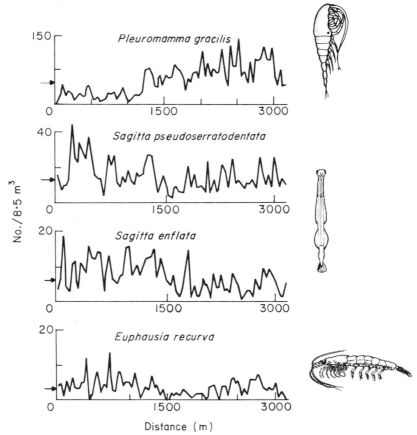

Most of these migrations form a similar pattern. During the daylight hours, the zooplankton are relatively deep in the water column, but during the evening dusk they ascend to the surface. They then disperse somewhat through the water column in the middle of the night, reaggregate at the surface before dawn, and then descend again to the day depth (by active downward swimming) as light intensity increases (Fig. 2.16). These movements correspond to those of the once enigmatic 'deep scattering layers' of the sea discovered during the Second World War and subsequently shown to comprise aggregations of migrating planktonic and nektonic animals.

Although many phylogenetically unrelated animals do show this or a similar pattern, it is by no means universal or stereotyped. It may be suppressed in certain life-history stages, at certain times, or in whole areas. In polar latitudes, for example, migrations are predominantly seasonal in that the zooplankton remain at the surface during the polar summer and at depth during the long, polar winter. Further, some species display (in some areas and/or at some times) reversed migrations, being near the surface during the day and at depth at night; other species migrate under some conditions but not under others, and many migrate to different day depths in different regions. A number of species do not carry out any vertical migrations at all.

Fig. 2.16 Characteristic vertical movements undertaken by zooplanktonic populations: (a) a surface-dwelling species such as the adult female of the copepod *Calanus*; and (b) a deeper-living species, e.g. an acanthephyrid prawn.

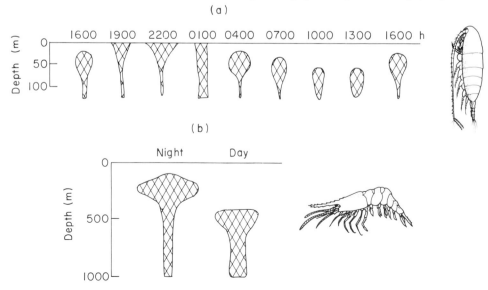

Interpretation of this widespread and important activity pattern has been bedevilled by failure to distinguish between causes, consequences and timing mechanisms, and very many hypotheses have been advanced in putative explanation. There seems little doubt that changes in light intensity provide the triggering and timing mechanisms for the various phases of the pattern: the zooplankton follow bands of constant light intensity as these bands move through the surface waters with regular day/night alternation. But there seems no reason to suppose that restriction to certain ambient light intensities provides the reason for the movements. Here many of the exceptions to the standard pattern are helpful in proving the rule. The exceptions listed above share the common factor of food availability (or lack thereof). The zooplankton remain in the immediately surface waters during the phytoplankton bloom of the polar summer, and remain at constant depth during the winter when no food is available and they are living on their food reserves. Stages in the life history which do not migrate are normally the non-feeding stages whereas the stages which show most marked vertical migrations are those requiring most food (e.g. adult females). Migratory movements are suppressed in the presence of abundant food and are stimulated in its absence. Non-migrants are either tied to a specific microhabitat (e.g. the neuston) or are predators. Therefore, as outlined on pp. 24–5 and as originally suggested by Hardy nearly thirty years ago, vertical migrations are most plausibly regarded as the only effective way in which an animal capable of limited lateral movement (relative to its water mass) may change its position in the surface waters when its immediate environment has reached near threshold concentrations of food.

Therefore, vertical movement on the part of the herbivores is a response to the horizontal patchiness of the phytoplankton (itself in part herbivore created), and the herbivores are followed by many of the carnivores which depend on them. Changes in light intensity provide a dependable timing mechanism and, all things once again being equal, it is probably selectively advantageous to move at such times as to keep within areas of relatively low light intensity in order to minimize the impact of visually hunting predators, i.e. to be at the surface at night and at depth during the day. In general, however, predators can be expected to be adapted to the prevailing environmental conditions in the habitats occupied by their prey.

Granted that these movements occur, many and varied will be the consequences. Populations will be well-mixed, increasing rates of gene flow and decreasing the incidence of speciation.

Consumption of the phytoplankton will decline at threshold concentrations providing the algae with an opportunity of recovery. Consumption of food in the relatively warm surface waters and its digestion at depth in cooler surroundings when vertically migrating may also lead to the gain of an 'energy bonus'. Metabolic rate and the energy required to sustain it are proportional to temperature and so activity is high in the surface waters, leading to the capture of many algae; but by migrating into cooler waters, less of the energy from the meal will have to be devoted to maintenance of a high metabolic rate (i.e. respired) and a greater proportion can then be put to growth, reproduction or storage. As we shall see later (p. 195), vertical migrations have been suggested as forming one route along which food materials pass into the deep sea.

2.6   The bacterioplankton

The pelagic zone abounds in various types of gram negative bacteria, many of them motile, which subsist on a diet of the dissolved organic compounds released by living phytoplankton and leached from dead tissues. Bacterioplanktonic productivity is very difficult to measure as any attempt to enclose them in small volumes of water or to culture them almost immediately alters the nature of the system. One series of samples taken from within the phytoplankton-maximum zone indicated that biomasses of some 6 mg C m^{-3} produced 42 mg C m^{-3} daily, but it is not yet known to what extent these values are representative or even realistic. Bacterial numbers appear to approximate 10^3 ml^{-1} of oceanic water, but higher densities are associated with the water adjacent to large masses of organic debris, for example that deriving from salt-marshes. More than 10^7 bacteria ml^{-1} and more than 10^5 of the largely bacteria-dependent microplankton ml^{-1} have been recorded from such microenvironments. Clearly, bacterially mediated food webs may prove to be much more important than their frequency of mention in reviews and standard texts might indicate: Sorokin (1971) estimated that in tropical waters, the production of bacterioplankton often exceeds that of the phytoplankton and forms just as important a food source.

2.7   Diversity and other community characteristics

The plankton of highly productive areas (zones of upwelling and some inshore regions) show clear differences to those characterizing open, unproductive tropical waters, and moderately productive systems are, as one might expect, intermediate in these respects. Several of the differences have already been mentioned, in passing, earlier in this chapter. Species in productive waters are typically large and inefficient and generate a large annual production in relation to their own biomass. The plankton as a whole is species-poor and has relatively little total community biomass supported by each unit of primary production. Carnivorous species are not well-represented, anti-consumer adaptations are not prevalent, and food chains may be relatively short and simple.

In contrast, planktonic species in unproductive waters are typically small, efficient, show a smaller annual production in relation to their biomass, and possess anti-consumer adaptations. Carnivores are an important element in the species-rich oceanic communities, food chains are long and relatively complex, and each unit of primary production supports a relatively large community biomass.

Most of these properties revolve around the efficiency which is at a premium when resources are in short supply relative to the needs of the consumers (though not necessarily in absolute terms), whether the consumers be phytoplankton or zooplankton. The high diversity of tropical oceanic plankton, however, may appear to be something of a paradox in that one might expect a highly competitive system to lead, through competitive exclusion, to species-poverty. This expectation rests on an assumption that conditions remain favourable to one particular competitor for a sufficient length of time to enable ousting of other species to occur, and this assumption is probably not justified in the pelagic environment. Neither does it allow for the evolution of dietary specialization. On a small scale the open ocean is not a uniform environment; changing patterns of turbulence, changing concentrations of inorganic and organic nutrients, etc., lead to a spatial and temporal mozaic of microhabitats, and balances of competitive advantage probably change continually. The pressure from consuming species, checking the potential build-up of any one prey species, may also have the effect of preventing a given species from monopolizing resources and achieving dominance, thereby acting as a counterbalance to any temporary superiority.

In these community characteristics, the oceanic/upwelling contrasts are almost exactly parallelled by other ecosystem dichotomies. Thus coral reefs, the deep sea, oligotrophic lakes and rain forest parallel the oceanic system, whilst mangrove-swamps, shallow lagoons, eutrophic lakes and boreal forests parallel the upwelling zones.

71

3 The Benthos of Continental Shelf and Littoral Sediments

3.1 Introduction

The pelagic environment is demonstrably three-dimensional and all three dimensions are exploited by its organisms; on turning to consider the benthos, however, we encounter essentially a two-dimensional habitat, superficially more akin to that of the land. It is not that the third dimension is absent—indeed continental shelf sediments may have depths to be measured in kilometres—but that, except for the surface film, marine sediments are anaerobic environments inhospitable to most organisms. Paradoxically, subsurface sediments are inhospitable by virtue of being rich in hydrogen sulphide, methane, ferrous ions, etc., which are formed as a result of biological activity. As organic materials settle out onto the sea bed and are incorporated into the substratum by further sedimentation, so they are decomposed by bacteria. In areas of plentiful oxygen, this gas can be used directly, but where the rate of supply is exceeded by the demand, respiration and fermentation can still proceed using oxidized inorganic and organic compounds as the hydrogen acceptors (i.e. in the reverse pathways to bacterial photo- and chemosynthesis). These anaerobic metabolic processes release the ammonia, hydrogen sulphide and other toxic reduced ions and molecules.

Between the areas of aerobic and anaerobic decomposition lies a zone of rapid transition in which, in particular, the redox potential (Eh) changes dramatically. This provides a useful diagnostic feature and the transitional zone is known as the 'redox discontinuity layer' (Fig. 3.1). Its depth within the sediment will depend on the quantity of organic matter available for decomposition and on the rate at which oxygen can diffuse down from the

Fig. 3.1 Changes in physical and chemical properties of sediments across the redox discontinuity layer (the 'grey layer'). (After Fenchel & Riedl 1970.)

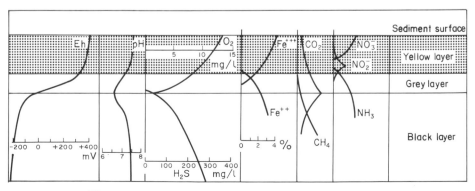

overlying water. Thus in organic muds, which are relatively impermeable to the oxygen-carrying water, the aerobic surface layer may only be one or two millimetres deep, whilst in permeable sands with a low rate of organic input aerobic conditions may extend down for several decimetres. Muds form in areas of low water velocity; sands have only particles of larger size because current or wind induced turbulence maintains the fine particles in suspension. Generally, therefore, sands occur in regions of rapid water movement, and the surface layers of the substratum may be suspended temporarily in 'underwater sand storms' enabling oxygen to penetrate deeply into the sea bed and providing a source of supply additional to permeation.

Some obligate and facultative anaerobes (which, respectively, require anoxia or are able to survive both oxic and anoxic conditions) are confined to the sulphurous redox discontinuity layer and there form a 'thiobiota' (see Fenchel & Riedl 1970 and p. 16), with a food web based on bacteria (especially on sulphate-reducing species such as *Desulfovibrio*). The meiofaunal metazoans show a number of characteristic features: prevalence of symbiotic bacteria; absence of mitochondria; very wide geographical distributions; and partitioning of the redox gradient between the various species. The difficulties of studying obligate anaerobes living within the sea bed are enormous, however, (they die, for example, if oxygen is allowed to come into contact with them) and as yet we know little of thiobiotic ecology.

Most organisms can only dwell below the redox discontinuity depth if they can oxygenate their immediate surroundings. The larger infaunal animals, for example, construct burrow-systems which open at the surface and, by drawing a current of water into or through the burrow, can create in its walls an extension of the surface sedimentary regime extending down up to one metre. The oxygen demand of the surrounding anoxic sediments continually depletes the burrow of oxygen and so the maintenance of an aerobic environment must be kept up throughout the life of the burrow-dweller. The water current, however, may also be used to bring a supply of suspended food to the animal and so may serve more than one function. By describing the occurrence of infaunal species, it is evident that the depth dimension is in fact exploited (albeit to only a few centimetres or decimetres at most), and even within this restricted range different species characteristically occupy different depths (Fig. 3.2). This is but one facet of the many competitive and interference interactions between the infaunal species brought about by the restricted living space (pp. 83–93).

73

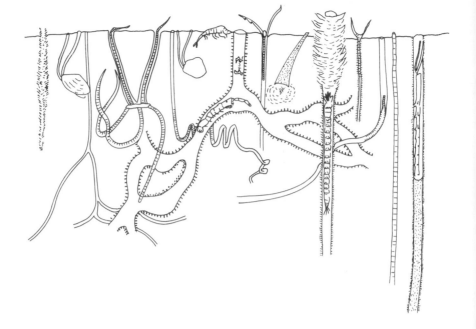

Fig. 3.2 Subdivision of the vertical gradient in sediments between a number of infaunal species: the position within the sediments of the dominant benthic animals of the shallow shelf sediments off Georgia, USA. (After Dörjes & Howard 1975.)

Infaunal species are of interest not only to biologists but also to geologists, and our knowledge of the daily lives of many benthic animals has been derived more from geological work than from the activities of biologists (see e.g. Schäfer 1972). For reasons which escape the authors, the study of palaeoecology is treated as a branch of geology, and in order to interpret palaeoecological data it is necessary to know how organisms become fossilized, with what biases and under what conditions. Hence organism-sediment relationships, such as burial and burrowing, have been studied for many years, especially by German 'actuo-palaeontologists'. Burrowing activities are also of interest to sedimentological geologists because bedding layers are disturbed as animals move through them (causing 'bioturbation'). Dense populations of polychaetes may circulate all the upper 10 cm of sediment through their guts in less than two years; and a population of the long, thin polychaete *Heteromastus* (8–15 cm long, 1 mm diameter) can move sediment from 10–30 cm depth to the surface during the four months of autumn at a rate of $2.5 \, \mathrm{l} \, \mathrm{m}^{-2}$ daily—this is equivalent to a layer 20 cm deep! For a long period, benthic biologists were preoccupied with descriptive and classificatory exercises (see pp. 98–101) and it is only relatively recently that ecological information rivalling in quality that obtained for geological purposes has become available (Fenchel (1978) and Gray (1981) provide good reviews).

3.2 Trophic relations

3.2.1 *Sources of food*

In essence there are two forms of input of organic matter into shelf/littoral benthic systems: detritus and living plankton from the overlying water mass; and living and non-living organic matter associated with the sediment itself. In addition, dissolved organics are certainly important in nourishing bacteria and they may be usable by some benthic animals although this is controversial and has been so for 70 years. If the quantity of living organisms beneath a given square metre of sea surface is accorded a value of unity, then that of particulate organic detritus in the water would lie in the range 1–10, and dissolved organics in the order of 50–500. The dissolved organics clearly constitute a vast pool of potential food (equivalent to the total production of the ocean for 30 years or more); much of the pool, however, appears to be highly resistant to metabolic processes, even those of bacteria, and it is probably effectively inert (as the large size of the pool might suggest). The current concensus (Jørgensen 1976) is that although dissolved organics may be of importance to some worm-like animals, they are unlikely to be generally of significance to metazoans.

The pelagic or benthic provenance of the primary food sources is reflected by the division of benthic organisms into suspension and deposit feeders respectively (p. 33). The division, as might be expected, is by no means absolute as the surface deposits may be put into suspension by water movements and thereby be made available to suspension feeders, and some organisms can feed in both modes. The division is nevertheless a useful one. With the exception of the microbenthic algae of littoral sediments, however, the truly benthic food sources are obtained, by sedimentation, from the water column— representing potential food materials which have not been intercepted by pelagic organisms. Hence it is most convenient to consider food inputs under two headings: microbenthic algae, and rates of fall-out onto the sea bed (regardless of whether they are taken by suspension or deposit feeders).

Unicellular and filamentous algae can only live on or in the surface layers of the sediment in the littoral zone, where the photic zone extends down to the substratum. There, productivities of the order of 0.2–1.3 g C m^{-2} daily can be achieved, dependent mainly on water clarity. Mat-forming species are confined to the relatively stable muds, whilst sands are characterized by small species living attached to, or between the grains. There,

75

benthic microalgae can photosynthesize over a much wider range of light intensity than can the planktonic species. Little or no photoinhibition occurs (Fig. 3.3), adapting littoral species to full sunlight at low tide, and they can also utilize very low intensities such as may be experienced at high tide.

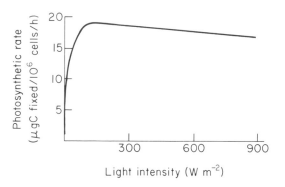

Fig. 3.3 The relationship between light intensity and photosynthetic production in benthic diatoms exposed to natural sunlight. (After Taylor 1964.) Note that full, midday sunlight (870 W m^{-2}) results in only a 10% inhibition of the maximum rate. Compare with Fig. 2.2.

Over the larger, continental shelf areas, this *in situ* photosynthesis is missing and the benthos is supported solely by pelagic production and by detrital materials emanating from the coastal fringe. The measurement of fall-out rate is difficult, not least because in order to estimate the quantity of material actually reaching the bottom, a cup or sediment collector must be sited near to the sea bed, but there it is also likely to collect material resuspended from the sediment surface, leading to an overestimation of deposition rate. Published rates are therefore variable within wide limits, but they do generally reflect: (a) the depth of water through which material must travel; (b) the magnitude of pelagic production; and (c) the proximity of additional sources of detritus. The relationship between pelagic photosynthesis and fall-out rate is not linear because a higher percentage of the production of nutrient-rich areas may reach a given depth than that in unproductive systems as a consequence of the relative inefficiency of utilization discussed earlier (p. 55 and Fig. 3.4). As an approximate indication of magnitude, the annual input of organic carbon onto the continental shelf probably averages some 100 g C m^{-2}, with somewhat less than 25 g C m^{-2} yearly near the edge of the shelf, and more than 300 g C m^{-2} yearly in favourable littoral areas. Analysis of the contents of sediment traps suggests that most of this input is in the form of the faecal pellets of herbivorous zooplankton, in which much material remains undigested (p. 66).

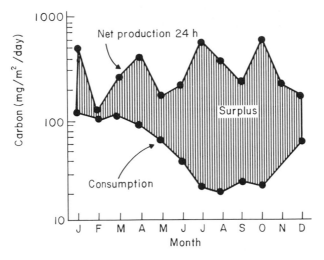

Fig. 3.4 The relationship between net phytoplanktonic production and zooplanktonic consumption in a lagoonal area of Kerala, India, showing a large, ungrazed 'surplus' of phytoplankton. (After Qasim 1970.)

The planktonic food available to suspension feeders is naturally related to the same three variables as is fall-out rate, and, in addition, to the extent of water movement near to the sea bed which can bring about renewal of suspended supplies. Suspension feeders may be of great importance in translocating food from the water column to the sediment surface. Much of the material removed from suspension may not be suitable food for the organism concerned or may not be digested; materials which are not ingested are rejected as 'pseudofaecal' pellets, often bound in mucus, and both the faeces and pseudofaeces are deposited on the substratum and are there available to deposit feeders. In oysters, for example, this biodeposition may amount to over 1 kg C m^{-2} yearly.

Budgets for two, shallow-water, benthic systems are given in Table 3.1 and these indicate the variation which occurs in the relative importance of different forms of food input.

3.2.2 Utilization of food sources

Microbenthic algae and dead phytoplankton cells may be consumed and digested by benthic animals as easily and as efficiently

Table 3.1 Inputs of food materials into the benthic systems of two, shallow-water, marine environments. (From Wolff 1977 and Day *et al.* 1973.)

Grevelingen Estuary, Netherlands (g C m^{-2} yearly)		
Detritus from fringing vegetation:		
salt-marshes		0.3–7
sea-grass meadows		5–30
Detritus from adjacent sea		155–225
Run-off from adjacent land		2
Phytoplankton		130
Benthic microflora		25–57
	Total	317–451
Barataria Bay, Louisiana (g C m^{-2} yearly)		
Detritus from fringing salt-marshes		297
Phytoplankton		209
Benthic microflora		244
	Total	750

as planktonic herbivores obtain their food, and the same is true of the living planktonic component of the diet of suspension feeders' This is not the case with the detrital component of benthic diets. If we were able to follow the passage of a piece of 'proto-detritus' as it slowly falls through the water column, we would see it loose soluble organic compounds through leaching and others would be taken out as it passed through several guts before finally arriving on the sea bed. By this time, it would probably contain only skeletal or other refractory substances and be of little direct food value to animals. In and on the sediments, however, it will be colonized by various types of bacterium and they will eventually support bacterial-feeding populations of protists and meiofaunal metazoans. Hence a unit of detritus, when it is ingested by, say, a deposit-feeding polychaete. will in fact be a microcosm in its own right, with bacteria, ciliates, amoebae, flatworms, nematodes, etc., all associated with the original detrital flake. In the littoral zone, the same particle would also provide a substratum for various blue-green algae and photosynthetic protists. The question therefore arises: What are the detritus feeders actually assimilating when they ingest this aggregate? The answer cannot be provided with absolute certainty, but the presumption that they are digesting the bacteria, protists and meiofauna is very strong.

The evidence for this presumption comes from several sources. By comparing the organic content of the deposits consumed and the faeces produced afterwards, it is possible to estimate the efficiency with which the food is assimilated. For many detritus-feeding species the assimilation efficiency is very low (1–10%) which indicates that most of the organic content of the food is indigestible. (Some species, however, appear able to select which items from within the matrix to ingest and in these, assimilation efficiencies are much higher—up to nearly 50%.) Yet when in the laboratory thesespecies are fed on diets of pure bacteria, for example, efficiency is greatly increased (to some 70%), and by sampling the ingesta from different regions of the gut it can be shown that the bacterial component decreases on passage, particularly in the region between stomach and intestine. When the relatively bacteria-free faecal material is released to the environment, an increasing microbial metabolism is observed through time until a value corresponding to the steady state system maintained in the rest of the sediment is achieved (see below). After this point, no further increase is noted until and unless the detrital material is rejuvenated by passage through the gut of another detritus feeder. Detritus feeders also appear enzymatically ill-equipped to cope with the refractory organic compounds of pure detritus which is anything other than fresh. Utilization of meiofauna by the deposit-feeding macrofauna is still under debate (see Kuipers *et al.* 1981). Circumstantial evidence for a nutritional role can be found in low meiofaunal densities in regions of high detritus-feeder abundance and vice versa, and in the fact that energy budgets calculated for benthic systems will not balance unless the macrofauna are consuming meiofaunal species (we will return to this topic in section 3.2.3.).

Such lines of evidence suggest that consumers of deposits are ingesting the sediment and its associated detrital materials (with various degrees of selectivity), are digesting the living components of the detrital aggregate, and are voiding the relatively inert non-living components back into the environment, where they can be recolonized by bacteria, protists and meiofauna and the cycle repeated. The production of the deposit and suspension feeders is then available to benthic carnivores (although, of course, the deposit and suspension feeders are themselves carnivorous to a large extent, as are many of the meiofauna) and to benthic-feeding members of the nekton.

3.2.3   *Production and biomass*

The productivity of suspension feeders will be determined by the rate of supply of their suspended food and, ultimately at least, that of the deposit feeders will be governed by the rate of fall-out of potential food particles. But some shallow-water sediments are so rich in deposited organics that 50% of the respiration of fixed organic matter may be accomplished in the anoxic regions and undecomposed tissues may be removed from circulation by burial—food may therefore appear superabundant in the surface deposits. As pointed out above, however, not all the organic content of marine sediments may immediately be available as food, and it may be bacterial production which is really fueling the system. What then may determine bacterial productivity? We saw earlier (section 2.4) that for every unit amount of carbon, bacteria have high demands for nitrogen and phosphorus and these they obtain from the surrounding water, rather than from the nutrient-poor detritus. Sediments, however, have low permeabilities and accordingly it appears likely that local nutrient shortage could limit bacterial production, not shortage of detrital carbon. It is here that the role of detritus feeders (or consumers of bacteria in general) is important. By consuming the bacteria, the various protists and animals release their incorporated nutrients back into circulation again and permit continued bacterial production: in effect, the consumers maintain the bacteria in growth phases. Movement of consumers within the sediment (and burrow irrigation, etc.) will also cause local turbulence of the interstitial water, resulting in replenished stocks of nutrients for the bacteria. Indeed in the last ten years it has been found that, contrary to earlier postulation, bacteria do not cover the entire surface of detrital particles; colonies are often small and in a static (no growth) phase. Predation may be, through the nutrient regeneration caused, the single most effective factor in maintaining the stimilating bacterial production. It has even been argued that, as both the meiofauna and macrofauna are dependent on bacteria for food, they do not compete for this resource since the meiofauna, by their movement and by recycling nutrients, induce a bacterial productivity which could not occur without them, and then subsist on that production. Some nematodes may also 'garden' microalgae; the latter multiplying in the mucus trails of the nematodes and later being consumed as these meiofaunal organisms retrace their movements.

A further limit on benthic production may arise from the cohesiveness of the faecal pellets produced by detritus feeders (Fig. 3.5). Although in effect the surface sediment is being recycled continually through animals' guts, some species are disinclined to ingest what can be recognized as freshly produced faecal pellets. (This would clearly be of selective advantage in that such pellets require an 'incubation time' before yielding digestible food (see p. 79) during which they also break down again to diffuse sediment.) Hence abundant populations of deposit feeders may render their own habitat temporarily unattractive, both reducing their own productivity and permitting the bacteria to decline to a steady, nutrient-limited state.

The effects of these limitations in determining general levels of abundance and productivity are poorly known. Unfortunately, early work on the distribution of benthic biomass (little attention then being paid to production) was often quoted in wet or fresh weights per unit surface area, and water and ash content vary markedly from one group of organisms to another. For the purposes of comparison with data for the deep sea given in Chapter 7, wet weight biomasses on the continental shelf generally lie within the range 150–500 g m^{-2}, with a maximum of over 1 kg m^{-2} on the rich Antarctic Shelf. Littoral biomasses are not dissimilar.

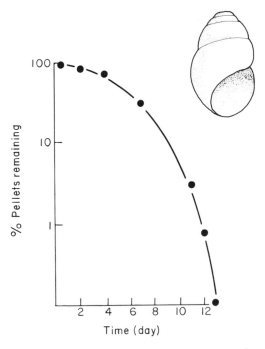

Fig. 3.5 The breakdown rate (into diffuse sediment) of the faecal pellets of a deposit-feeding gastropod, *Hydrobia minuta*. (After Levinton & Lopez 1977.)

Only relatively recently have productivities been measured (as opposed to inferred) and biomasses quoted as ash-free dry weights. Gerlach (1978) provides average figures for the abundance of the various size groups of organisms comprising the infauna of the shallow shelf: deposit feeders (120 μg ml^{-1}); other macrofauna (80 μg ml^{-1}); meiofauna (50 μg ml^{-1}); protists

$(20 \ \mu g \ ml^{-1})$; bacteria $(100 \ \mu g \ ml^{-1})$. He also estimated that up to 105 bacterial duplications per year would be required to support this system. On an areal basis, biomasses of shelf macrofauna recorded to date range from some 50 g (ash-free dry wt) m^{-2} down to less than 4 g m^{-2}, with the highest biomasses in inshore waters. In these same waters, annual macrofaunal production is only one or two times their average annual biomass, decreasing even further to less than half the average biomass in deeper areas. The smaller meiofauna probably produce annually more than five times, and sometimes more than ten times, their standing stock. As one might expect, the secondary production in organic muds and muddy sands, and in areas supporting dense beds of suspension feeders, is usually larger than that supported by clean sands both in absolute terms and as a rate of turnover of biomass. Even so, the production of the benthic macrofauna rarely exceeds 5% of that in the overlying water and their communities are often slow growing and long lived.

3.2.4 *Interactions between plankton and benthos*

Many of the macrobenthic species of littoral and shelf regions produce larvae which swim and feed in the pelagic system for about four weeks and in this manner by-pass the losses of potential food inherent in fccding on phytoplanktonic production only after it has sunk to the sea bed. The efficacy of this strategy will vary with latitude, since the pattern of primary production shows latitudinal differences (pp. 36–7). Production is continuous over the tropical shelves and here some 80–85% of benthic species have planktotrophic larvae; in temperate regions, phytoplanktonic production is seasonal and less predictable in its duration, and the percentage with planktotrophic larvae falls to near 60%; and in the polar zones, where photosynthesis is possible only for a few weeks, such larval types do not occur. A further 15% of tropical and temperate species, and 5% of those near the poles, produce realtively short-lived larvae which form part of the plankton but do not feed there at all or only to a vary minor extent. For the latter, pelagic life serves as a mechanism for short-distance dispersal, and, of course, dispersal will also result from the possession of planktotrophic larvae (see Chapter 9). However, if larval life in the plankton is to be interpreted in part as a means of occupying a habitat in which food is relatively abundant, it is legitimate to question why the pelagic larvae should ever leave these areas of plenty and return to the comparatively food-poor benthos. Obviously it is difficult to conceive of a pelagic

or clam, but the answer to this conundrum may be that as many as could manage to achieve permanently pelagic life have done so. Many authors have suggested that animal life evolved within the benthos and that the individual groups comprising the zooplankton (in both fresh water and the sea) have originated by paedomorphosis or neoteny from the larval stages of benthic species (see De Beer 1958; Gould 1977).

Passage of elements in the food web is never entirely one way, however. We noted earlier (p. 66) that pelagic animals may be important in releasing the nutrients required by the phytoplankton and returning them to their dissolved inorganic state in the water. Such a role is also performed by the benthos, as thefollowing example will illustrate. Nixon *et al.* (1976) investigated the regeneration of nitrogen by three, different, shallow-water, benthic systems in Narragansett Bay, Rhode Island, and found that the release of ammonia by benthic animals was responsible for the seasonal pattern of ammonia availability in the pelagic zone. Not all the nitrogen regenerated was in the form of ammonia, however, although nitrite and nitrate release were insignificant. A ratio of oxygen to nitrogen atoms of some 13 : 1 is usually required for the respiration of organic matter, but in Narragansett Bay the ratio of oxygen consumed by the benthos to inorganic nitrogen production was near to 30 : 1. Less than half of the expected nitrogen was being released not only in relation to the oxygen consumed but also to the quantity of phosphate regenerated. It appeared most likely that the 'missing' nitrogen was in the dissolved organic form which would not be available directly to most phytoplankton species. This observation may in part account for the fact that nitrogen, rather than phosphorus, may be the nutrient most often likely to limit pelagic photosynthesis.

3.3 Intra- and interspecific interactions

3.3.1 The nature of benthic interactions

Because the volume of sediment above the redox discontinuity layer is small and because bacterial productivity may be limited by unfavourable nutrient conditions within the substratum, the habitat of the benthos is such as to render competition for space and/or food likely. Whether or not competition is a common occurrence, if so between which types of species, and whether its incidence is reduced by predation, have all been the subject of copious theoretical argument (much of it summarized by Gray

1981) but as yet little experimental investigation or analysis. Indeed Gray concluded his discussion by commenting, 'We have done enough theorizing for the time being and more concrete data of a natural history type are needed before any material advances in our understanding of factors controlling the structure of benthic communities will be reached' (one might add that this sentiment is by no means true only of benthic ecology). Part of the difficulty is that competition need not be reflected by mortality: slower growth rates, high emigration rates and less energy expenditure on reproduction are all alternative reactions and they are not so easy to demonstrate or measure. Before proceeding, however, it is necessary to define competition. Competition may occur when any resource is shared, but it will *only* result when and where the quantity of some resource is insufficient for the needs of all the interested consumers. Resource sharing is a necessary but not a sufficient requirement for competition—a fact often overlooked.

One of the best known interactions between benthic organisms is where one population renders the habitat unsuitable for another. This may be achieved by means of chemical conditioning, as when the gastrotrich *Turbanella* releases a substance into the surrounding sediments which leads to avoidance of them by the potentially competing polychaete *Protodriloides*. More often (although this may merely reflect the greater ease with which it can be demonstrated), the effect may be an 'accidental' result of bioturbation, pelletization of the sediment or any other alteration of a sediment's properties by the locomotory or feeding activities of an organism. This is usually termed 'interference competition' although it has never been demonstrated to fall within the definition of competition given above. No resource necessarily reaches a limiting state; rather, one group of organisms—or even one individual—can be considered to pre-empt a resource which *might* become limiting at some time in the future. Other instances of interference competition are really more in the nature of accidental or indiscriminate predation or mortality. Suspension feeders and deposit feeders which vacuum-clean the surface sediment, for example, may prevent the larval stages of other species from settling on and colonizing a given area by consuming them or incorporating them in pseudofaecal pellets.

In the following sections we will examine these various biological interactions which can control the numbers of individuals per unit area and can influence the species composition of local patches of sediment. The relative scarcity of studies, and Gray's injunction above, will not permit many general hypotheses to be advanced; the accounts will therefore be based on a rather limited range of examples chosen to illustrate what may prove to be widespread phenomena. Considering all the cases of interactions to be presented below, one is left with the impression that much more remains to be learned and that many more detailed examples must be forthcoming before any accurate generalizations can emerge. The field, however, is a rapidly growing one.

3.3.2   *Interference competition*

Various authors have claimed that there are at least four main types of benthic organism whose life styles will cause them to interact: those that feed by collecting material as it settles on to the sediment surface; those that construct tubes or mucus-lined burrows or otherwise pre-empt and stabilize the sediment; those that move actively through the substratum; and those that produce large quantities of faecal and pseudofaecal materials.

Monospecific stands of epifaunal or shallowly infaunal brittlestars are not infrequent on many areas of the continental shelf (Fig. 3.6). Brittlestars are omnivorous and dense aggregations

Fig. 3.6 An aggregation of the brittlestar *Ophiothrix fragilis* on the sea bed. (Photograph and copyright G. F. Warner.)

appear able to consume almost anything which descends on to them from out of the plankton, including metamorphosing larval stages. Only for a brief period whilst they brood their young does their feeding cease and this provides the only opportunity for other species to invade the community and to grow sufficiently large to be beyond the size range of brittlestar prey before the latter resume their feeding. A similar effect may be responsible for some year-class phenomena in bivalves. A successful spatfall on to an area of sediment may lead to a strong year class of say, *Scrobicularia*, which can vacuum-clean the whole of the surface of the area colonized. If sufficient of the animals survive through to the next breeding season, they may destroy larvae of their own as well as of other species, and not until that year class loses its stronghold on the area can other year classes or other species invade. The small-scale distribution pattern of *Scrobicularia* may therefore be a patchwork of different age groups, with some of the patches comprising the oldest individuals which will die before the next spatfall and provide the potential settlement sites for the next generation of larvae. Year classes of bivalves always vary considerably in the success with which they can establish themselves, however, and so in any one area different year classes will be variably represented. In the cases of both the brittlestars and the bivalves, interference is indiscriminate as to the type of organism with which their feeding mechanisms interfere, except that only organisms seeking to colonize from above are affected. This does not mean that no other species can coexist. The poly-chaetes *Arenicola* and *Hediste* both occur with *Scrobicularia*, for example, because their larval life can be passed within the par-ental burrow-system and dispersal can take place at a size beyond that which can be caught by the bivalve.

Deposit feeders have been suggested to interact with suspen-sion feeders. The deposit feeders by continually reworking the sediment, aggregating it into faecal pellets (90% of some sedi-ments is in the form of discrete pellets), and rendering it more liquid may prevent the successful settlement of suspension feeders which require a stable, firm substratum; whilst dense beds of suspension feeders may prevent colonization by the larvae of deposit feeders as described above. Similarly, interac-tion between tube-dwellers and burrowers has been described by Woodin (1974). By excluding the tube-dwelling polychaete *Platynereis*, she demonstrated a three-fold increase in numbers of the burrowing polychaete *Armandia* and suggested that *Platy-nereis* monopolized the available space and interfered with the activity patterns of the burrower. There are, however, many cockle examples of sediments in which deposit feeders and suspension feeders coexist and in which tube-dwellers occur together with burrowers, even on microscales of the order of one hundredth of a square metre. Moreover, since animals of the same type are likely to have similar requirements, one might expect that competition would be most intense within a feeding type rather than between completely different general types. Evidence for this expectation is not difficult to find. The deeply burrowing bivalve *Yoldia*, for example, is known to disrupt the more shallow burrows of another bivalve, *Solemya*, when the two are maintained together in aquaria (Fig. 3.7). Both deposit feeders and suspension feeders also often partition the depth and other gradients between a range of similar species and feed or live at different depths or in different areas. The form taken by the presumed competition which leads to such stratification is not known, although it would appear to be an active process in that when a given species is absent or is removed from its station, another may expand its range to include the vacant horizon.

One may therefore find a whole range of potential interactions in nature—from suspension feeders rendering a habitat unsuitable for mobile deposit feeders (another example is given on p. 93), through no overt interaction, to cases in which suspension feeders such as oysters and mussels create conditions favourable

Fig. 3.7 Interactions within the sediment between the bivalves *Yoldia limatula* (YM), *Solemya velum* (SV) and *Nucula proxima* (NP). *Solemya* detours when confronted by *Yoldia*. The bar represents 1 cm. (After Levinton 1977.)

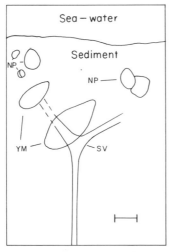

August 17 1969, 4 p.m.

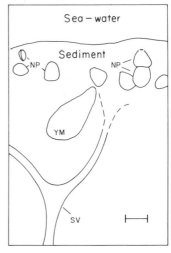

August 19 1969, 10 a.m.

to deposit feeders by depositing large quantities of silty faeces and pseudofaeces onto what may originally have been a hard substratum. The bivalves maintain themselves above the muddy sediment created by attaching to the shells of previous generations of their own species buried in the sediment. Equally, although some species may avoid pelletized sediment (p. 80), others appear to feed solely on faecal pellets. Few generalizations seem possible and interactions between individual pairs of species are almost irrespective of feeding type or life style. Whilst those taking place within a single feeding type are likely to be directly of selective advantage to the interferer, those occurring between completely different types of organism are likely to include a number which can only be described as accidental.

3.3.3 Predation

The effect of predators on the abundance and composition of the benthic fauna is, as elsewhere (pp. 57–8), somewhat variable and problematic. Woodin's findings mentioned above were obtained by enclosing areas of sediment within a cage preventing access by various species, and several authors have used the same technique to study the effect of predator exclusion (Peterson 1979). Three results appear general: (1) the densities of individuals within the cages that exclude predators are larger than in controls; (2) more infaunal species are present in the absence of predators; and (3) no marked increase in dominance by any single species appears to correlate with decrease in predation mortality. The two latter contrast markedly with equivalent studies which have been carried out on rocky shores (pp. 130 *et seq.*). These have shown that predation, by maintaining the numbers of potentially superior competitors at low levels, prevents dominance by one or a few species and the competitive exclusion that results, and thereby maintains high species diversities. In the absence of predation on rocky shores, the number of species declines and dominance by a few species increases, in at least some components of the fauna (these studies have rarely investigated the whole fauna however). One could argue from the results of caging experiments on soft sediments that predators normally keep both density and species richness below carrying capacity and that the larger benthic species are therefore not food but predator limited (Peterson 1979; Holland *et al.* 1980). When predation is removed, one might expect competition to increase but eventual competitive exclusion in the absence of predators might take a long time—longer than most caging experiments have lasted—and the more immediate effect of interspecific competition might in any event be reduced growth rates not exclusion. On the other hand, the act of installing a cage creates not only a disturbance but also alters the conditions inside it (increasing sedimentation, decreasing rates of water flow, etc.), and it has been argued (Gray 1981) that the increase in numbers of individuals and of species inside the cages are to some extent caused by colonization of the caged habitats by opportunistic, r-selected species absent from nearby control areas.

The effect of predators may not be nearly so clear-cut as the cage results seem to imply and Peterson's conclusions do not necessarily hold for all elements of the fauna; discounting the cage artifacts, only a few species often appear to be affected by the absence of external predators, the infauna least of all. This may be for a relatively simple reason. Predators moving over the surface of the sediment (i.e. those excluded by the cages) can often only obtain part of any infaunal prey individual: predators bite off those parts of the larger species projecting near or above the surface. The prey can regenerate the missing biomass and in some cases the missing organs. *Arenicola*, for example, may lose the tip of its tail to wading birds, fish and *Hediste* (and peacock worms lose their tentacular crowns to flatfish). *Arenicola* is not killed by this indignity and although it cannot regenerate the missing segments, it can increase its tail biomass back to the original value by increasing the length of those segments remaining. Thus energy has to be diverted from growth in general to regeneration, and the effect of predation is to reduce *Arenicola* productivity but not necessarily to decrease lugworm numbers: one third of the mean annual biomass of *Arenicola* may be removed by flatfish, etc., in the form of tail tips. The bivalve *Tellina*, however, may be prevented from breeding by the cropping of its siphons by young plaice. Epifaunal species are, of course, more exposed to predatory attack and often bear specific antipredator devices such as thick shells, spicular skeletons, fast escape responses, etc.; whilst amongst the infauna it is the deeper-burrowing species that are most protected.

3.3.4 *Competition for resources*

The evidence for direct intra- and interspecific competition is often circumstantial; nevertheless Levinton (1979), in particular, has argued strongly that it must occur. One could almost argue from the above section that in cases where predation is not a factor likely to control benthic densities, competition must be the process preventing unlimited expansion, although direct effects of the physical and biological environment are certainly known in several areas and can even cause complete defaunation (see pp. 94–5). Such 'climatic' control is not the case over most of the continental shelf however. Levinton has argued that it should be deposit feeders which compete most for food, suspension feeders competing mainly for space. As in section 3.3.2, however, the available data suggest few general rules. Competition for food between the suspension feeding *Crepidula* and *Ostrea* or *Crassostrea*, for example, has led to expensive attempts to remove *Crepidula* from oyster beds; and the densities of the larvae of both suspension feeders (such as cockles) and deposit feeders (such as ragworms) which settle out of the plankton are often much more than the available food or space could support were they all to grow to adult size. In many cases, the evidence for competitive interaction is very strong (Fig. 3.8; see also Chapter 9) but the nature of the interaction is not known.

Most detailed work has concerned the deposit-feeding gastropods and a number of studies have shown that in shallow waters these surface-living snails can graze down the microbenthic algae to levels which would be expected to induce both intra- and interspecific competition for food. *Hydrobia*, for example, eliminated diatoms from patches of sediment which had originally supported 5×10^5 diatoms per cm^2. It was able to home-in on rich patches by behavioural means, detecting the difference between areas of sediment artificially enriched with diatoms and control areas of the natural substratum. Similarly, *Nassarius* has been shown to remove more than 10% of the surface chlorophyll per day. When these snails were removed from experimental areas of their respective natural habitats, dramatic increases in algal numbers occurred.

More is known of the biology of the hydrobiid snails than of any other deposit feeder and Fenchel (e.g. 1975) has studied their competitive interactions in great detail. He investigated the ecology of *Hydrobia ulvae* and *H. ventrosa* in regions where they were occurring together (sympatric) and where each was living separately (allopatric). In allopatric populations, the two species are very similar to each other in most respects, but in localities where they are found together Fenchel found that they diverged from each other both morphologically and in their ecology, i.e. they showed 'character displacement'. When sympatric, *H. ulvae* is slightly larger and *H. ventrosa* much smaller than normal, so that there is a length ratio of 1.3 : 1 (or a weight ratio of 2 : 1) between them. These ratios are frequently found in closely-related

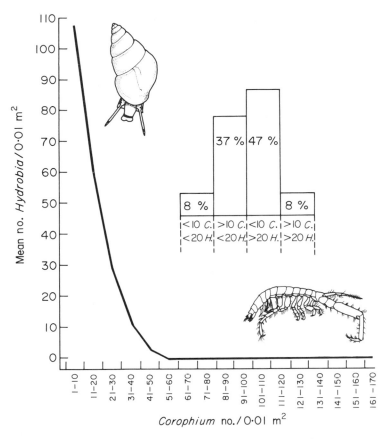

Fig. 3.8 Average numbers of the gastropod *Hydrobia ulvae* found in the same 0.01 m² area as differing densities of the amphipod *Corophium arenarium*. The inset histogram displays the proportion of samples (out of 315 total) with different relative abundances of these two, deposit-feeding species. The numbers of the two are clearly inversely correlated. (Barnes, unpublished data from Norfolk, UK.)

pairs of coexisting species and Hutchinson has demonstrated that they reflect the minimum size-difference necessary to enable the exploitation of different resources, and thereby prevent competitive exclusion. Other examples of character displacement shown by these two species when sympatric are that their reproductive periods become shorter and scarcely overlap and, largely as a direct consequence of their size difference, the size of food particles which they ingest diverges (Fig. 3.9). The exact nature of the competitive process was not elucidated, but the observation that their food differs when occurring sympatrically does clearly suggest a form of competition for food, and the same may hold true

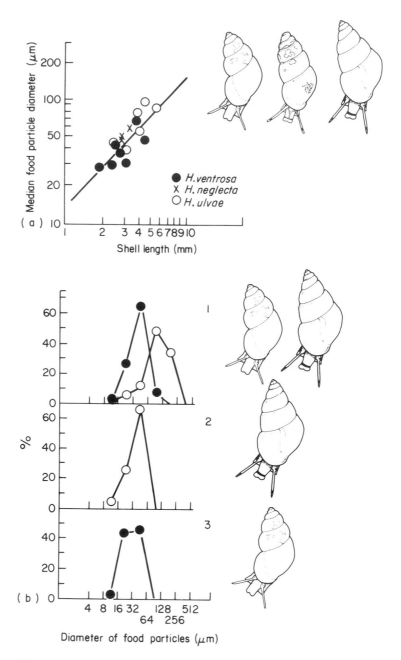

Fig. 3.9 The size of food particles taken by *Hydrobia* spp. in Danish localities. (a) The size of the food particle in relation to the size of the gastropod (note that size for size the three species consume the same size of particle); (b) the size of the food particle consumed by *H. ulvae* (open circles) and *H. ventrosa* (solid circles) when occurring together (1) and separately (2, 3). (After Fenchel 1975.)

intraspecifically when they occur allopatrically. Over most of their ranges *H. ulvae* appears the superior competitor and dominance by *H. ventrosa* is restricted to those few areas, such as regions of low salinity, in which *H. ulvae* is at a disadvantage, and to those sites to which *H. ulvae* has not been able to gain access, such as some coastal lagoons (see Barnes 1980a).

3.3.5   Stability

Many benthic assemblages of species seem fairly stable in the long-term but several fluctuate in the short-term. In temperate latitudes, seasonal cycles of recruitment occur with larval settlement in summer and heavy mortality during the winter (Fig. 3.10), imparting an annual variation in abundance. In a number of areas, oscillating patterns of dominance over a variety of time scales have been demonstrated. In Barnstaple Harbor, Maine, for example, Mills (1969) has documented the seasonal displacement of a deposit-feeding snail by a tube-dwelling amphipod as a result of interference competition, followed in turn by displacement of the amphipod by environmental effects. During the winter, the benthos of Barnstaple Harbor is dominated by the gastropod *Nassarius*, but in spring an amphipod *Ampelisca* colonizes the sediment and rapidly builds up its population density. The tubes of *Ampelisca* interfere with the freedom of movement and potential ingestion rate of the snail, and its population declines in inverse correlation to the rise in that of the amphipod.

Fig. 3.10 Annual fluctuations in the density and biomass of benthic macrofauna off the Northumberland coast, UK. (After Buchanan *et al.* 1978.)

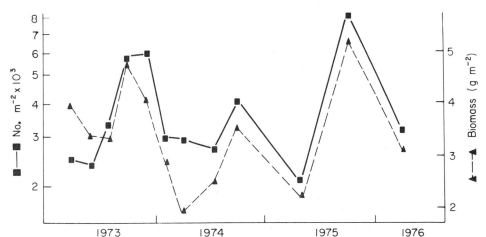

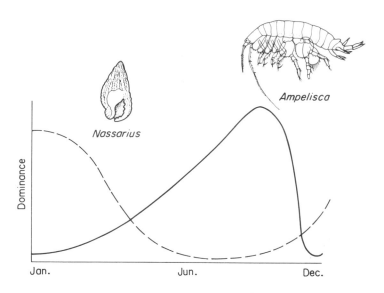

Fig. 3.11 An oscillating system of dominance between the tube-building amphipod *Ampelisca* and the deposit-feeding gastropod *Nassarius*. (After Mills 1969.)

In the autumn, however, storms extend their influence to the sea bed and the massed tubes of *Ampelisca* are physically 'rolled up like a carpet', freeing the sediment for recolonization by *Nassarius* (Fig. 3.11). An oscillating cycle with a longer period, some six to seven years, has been described in the Baltic Sea. Here another amphipod *Pontoporeia* alternates with the bivalve *Macoma* in a system also involving interference competition, this time interference with the larval recruitment of the bivalve.

A few areas of sea bed are subject to marked instability, however; sediments may be removed by storm action and deposited elsewhere, and intertidal sediments in particular may pass through cycles of erosion and accretion. Benthic organisms may therefore suffer periodic mass mortality as their habitat disappears or is buried beneath deposited material. Mass mortality may also have a biological cause as in the blooms of toxic dinoflagellates which afflict the coast of Florida amongst other areas. In the summer of 1971 a bloom of *Gymnodinium breve* killed 97% of the macrobenthos of Old Tampa Bay, Florida (removing 77% of the infaunal species). The pattern of recovery was studied by Simon and Dauer (1977). They found that the rate of recolonization was proportional to the dispersal faculties of the fauna, with an initial rapid influx of opportunistic polychaetes (Fig. 3.12).

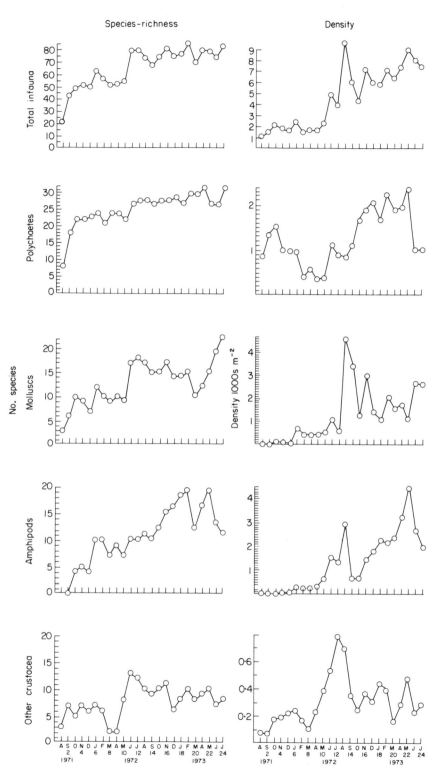

Fig. 3.12 The pattern of increase in species-richness and density per unit area of different elements in the macrofauna of a benthic system after natural defaunation. (After Simon & Dauer 1977.)

After a year, the fauna was similar to that present before the bloom but differences of detail were still present two years later. Comparable events occurred in Britain during and after the severe winter of 1962/1963: eighteen years later, some species have still not recovered their former range or abundance.

3.4 Diversity

In Chapter 1 it was pointed out that the global pattern of species-richness over the continental shelf was determined almost solely by that of the epifauna (p. 40), the tropical benthos being species rich and diversity declining polewards, and we can now attempt to analyse this pattern. We must start, however, by stressing that the factors responsible for the generation of high diversity are not necessarily those responsible for its maintenance. The creation of diversity is primarily a long-term process, being the result of speciation. This will be considered in detail in Chapter 10 but we can anticipate a little. A glance at a world map will suffice to show that the tropical areas of ocean are isolated from each other by north–south aligned continents and that within the rich Indo-West Pacific, shallow waters are further isolated by occurring in the form of rings around the numerous Malaysian, Indonesian and Philippino islands. There are less potential geographical barriers in higher latitudes and, if for no other reason, speciation may therefore be expected to be highest in shallow, tropical waters.

The questions for consideration here are: How have so many epifaunal species managed to coexist in the tropics without competitive exclusion? and Why is the change in species richness manifest only in the epifauna? At one level, the answer to the first question may reside in the observation that different, closely-related species have specialized by utilizing different sections of the resource spectrum, and it can also be noted that many such species do not in fact coexist, but replace each other geographically; because many species have only limited distributions, great numbers may be present in a single, large region. However, we are essentially concerned with numbers of species which do coexist and here demonstrations of character-displacement (p. 92) only push the question one stage further back. Why, for example, did character displacement evolve rather than competitive exclusion? Benthic ecology has provided the inspiration for several general theories of the generation and maintenance of diversity (especially that of Sanders (1968) and those stressing the importance of predation) but the theory which most successfully accounts for the observed distribution did not have a marine ancestry. Huston (1979) related maintenance of diversity to the interaction of two population parameters: potential rates of population increase and the frequency of population reduction (by predators, adverse environmental conditions or any other factors causing mortality). High rates of population growth in the fauna as a whole can lead to interspecific competition with the consequent possibility of competitive exclusion; high rates of population reduction occurring in the same system will reduce these likelihoods, although equivalently high mortalities may eliminate species populations unable to make good the loss. Diversity will, Huston's theory predicts, be highest when disturbance is just frequent enough to prevent competitive exclusion (see Table 3.2).

Table 3.2 Levels of diversity associated with combinations of population growth rates and frequencies of population reduction. (From data in Huston 1979.)

| | | Population growth rate | | | |
		Low	Moderate	High	Very High
Frequency	Low	Moderately low	Moderate	Low	Very low
of	Moderate	Moderate	Very high	Moderate	Low
population	High	Low	Moderate	High	Moderate
reduction	Very high	Very low	Low	Moderate	Moderate

As stressed above, the diversity apparent at any one point is a product of both the evolutionary process of species generation (coupled with subsequent biogeographic spread) and the ecological process of diversity maintenance, and it is not often easy to separate past and present events. Nevertheless, whilst bearing in mind the probability that the main centre of the formation of new species is in the tropics, it is possible to examine what is known of characteristic population growth rates in different latitudes and of the distribution of factors perturbing population growth in the light of Huston's thesis. Let us begin with the largely suspension-feeding epifauna.

Population growth rates are a function of the reproductive strategy of the species concerned and this in turn is related to the pattern of relative food abundance. In high latitudes, food is only available for a few months of the year and for the remaining period animals must subsist on food reserves. Species with planktonic larvae are rare (p. 82); most produce few relatively well-provisioned young, and population growth rates are low. In the tropics, food is available throughout the year but many mouths are interested in unit quantity of this food and growth rates are characteristically moderate to high. The seasonally abundant food resources of temperate zones with the equivalent 'make-hay-while-the-sun-shines' strategy of the epifauna results in as high or even higher population growth rates.

The latitudinal patterns of mortality factors are less easy to assess: predation intensity is probably highest in the tropics whilst environmentally induced stress and mortality, although high in several tropical areas, are likely to be most severe in high latitudes. We may therefore regard the frequency of population decrease as being relatively uniform latitudinally; if anything, since the high latitude epifauna are presumably adapted to the stressful environmental conditions prevailing, the rate of population decrease is likely to be least in high latitudes. Let us score this population parameter as being moderate.

The moderate rates of both population increase and decrease in the tropics would then lead to a species-rich epifauna being maintained; whilst the low rate of speciation and the low potential rates of population growth in high latitudes will result in species poverty. In temperate latitudes with their seasonally fluctuating climates, population growth rates may be high whilst speciation rates are low to moderate, yielding a moderately diverse fauna. In the tropics, therefore, the high predation rates coupled with the occasional period of environmental stress (the effects of hurricanes on mangrove-swamps and coral reefs, etc.) serve to maintain high diversities; and, all other things being equal, the fact that some of the dominant epifaunal organisms provide structures of great spatial heterogeneity (see p. 40)—structures absent from more temperate climes—permit even higher diversities to be supported by providing additional opportunities for microhabitat specialization.

Conditions are far more uniform for the infauna. Environmental fluctuations are damped within the sediments whatever the latitude. Nutrient limitations of bacterial productivity will also occur regardless of climatic regime with consequent effects on the reproductive strategies of deposit feeders; and all infaunal species are shielded by the depth of sediment between themselves and epifaunal consumers. Seasonal fluctuations in pelagic production are also damped out since pelagic material is only consumed after it has been incorporated into the large organic pool in the sediments. One might therefore predict that latitudinal variations would not markedly affect the infauna, and since interspecific competition for food is generally regarded as being more intense than in the epifauna, whilst other agencies effecting mortality are probably less intense, the diversity of the infauna will generally be lower than that achievable by the epifauna. Although we are still ignorant of the detailed population processes of benthic animals, Huston's argument does seem capable of explaining the marine—and other—patterns of diversity (see Fig. 1.22).

3.5 Benthic communities

Early work on the nature of the organisms inhabiting the sea bed identified a series of 'communities'—regular and characteristic associations of macrofaunal species—each named after the one or two species regarded as being most indicative and characteristic of that particular association. In this manner, large areas of the continental shelf were labelled as supporting various named communities; parallel communities were identified in geographically separate areas with similar environments and different variants or subcommunities were described in local regions. These developments paralleled the phytosociological approach to botanical ecology and, similarly, precise rules were laid down for the naming and characterization of communities, subcommunities, etc. (At the same time, a similar approach was also applied to pelagic ecology and various 'indicator species', indicative of different water masses, were proposed.)

Thus Thorson (1957) diagnosed dozens of 'communities' grouped into seven major types. *Macoma* communities dominated sheltered waters down to 60 m; *Tellina* communities occurred in shallow, sandy bottoms; *Venus* communities in deeper, but still sandy substrata (7–40 m); *Abra* communities typified the organic muds of estuaries and other sheltered regions; *Amphiura* communities were characteristic of soft sediments at greater depth (15–100 m); *Maldane/Ophiura* communities replaced those of *Abra* in the soft muds of the continental shelf down to 300 m; and a group of amphipod-dominated communities (including those of *Pontoporeia* and *Ampelisca* mentioned above) were found in localized areas of sheltered, muddy deposits (see Fig. 3.13). *Macoma* communities, for example, included that of *M. calcarea* around the Arctic Ocean, of *M. incongrua* in the north-west Pacific, of *M. nasuta* and *M. secta* in the north-east Pacific, and of *M. balthica* in the boreal north-east Atlantic; whilst the *Macoma balthica* community included variants dominated by cockles, lugworms, *Mya* or *Scrobicularia*.

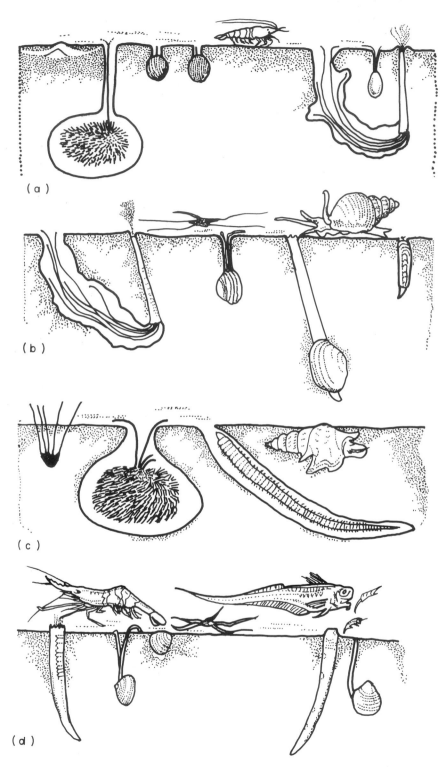

Fig. 3.13 Four benthic shelf 'communities': (a) a *Venus* community; (b) an *Abra* community; (c) an *Amphiura* community; and (d) a relatively deep water community. (After Thorson 1971.)

Although the names of many of these communities are still used for descriptive purposes—in much the same manner as one might talk of oak woodland or heather moor—their existence as discrete entities has fallen completely from favour. This has resulted from the application of more objective, numerical methods of analysis in benthic ecology which have disclosed that the subjective 'typical' assemblages of species characterizing the various 'communities' each graded smoothly into the others. Intermediates are the rule rather than the exception, and each benthic species appears to be distributed in relation to an individual series of requirements rather than as one of many species sharing a common response to any given environmental gradient. In many respects, the whole community story is equivalent to the impression which may be gained of the intertidal zone during marine field-courses. The latter often take students to 'typical' sandy shores, 'typical' mud-flats, 'typical' rocky shores, and so on, and it is easy to imagine that the intertidal zone comprises a rather limited series of discrete and distinguishable habitat types. If, however, one carries out a transect along a stretch of coastline, it is evident that habitat types grade into each other and that very few areas are uniform and 'typical'. Sharp boundaries, where they occur, are due not to interactions between different assemblages of species but to major, sudden changes in the environment, as when an area of rock outcrops on an otherwise sandy beach. In like manner, indicator species may characterize certain environmental regimes but they do not necessarily indicate the existence of a predictable suite of associated species.

This is not to suggest that several species do not occur together regularly, and interact with each other; clearly this is often the case. However, each local area of sediment supports an assemblage of species which differs in minor, and sometimes major, respects from all other such local areas, in part for historical reasons, in part as a result of biogeographical processes, in part because of ecological interactions *in situ*, and in part as a consequence of minor differences in the environment. Each local area therefore supports its own community at any one time, and a limited number of stereotyped 'communities' do not occur in the sea, except in very broad terms.

4 Salt-marshes, Mangrove-swamps and Sea-grass Meadows

This short chapter will be devoted to the role of the stands of vegetation of terrestrial ancestry which fringe the sea in sandy or muddy areas and function as coastal food factories. Their ecological interest is very much wider than we can cover here; their semi-terrestrial nature places them on the periphery of the main themes of the book and for other details the reader is referred to Reimold and Queen (1974), Chapman (1977), Ranwell (1972), McRoy and Helfferich (1977) and Phillips and McRoy (1980).

Sea-grasses are distributed over most of the globe wherever there is shallow water (except in high polar regions), but salt-marsh and mangrove-swamp replace each other geographically: salt-marshes typify cooler and/or drier coasts and mangrove-swamps hot, wet areas (Fig. 4.1). Only in the Gulf of Mexico states of the USA and in south-eastern Australia are both vegetation types present along the same stretch of coastline, although in dry regions within the tropical mangrove belt a form of salt-marsh can replace the mangrove flora on some parts of the shore. A major difference between salt-marshes and mangrove-swamps

Fig. 4.1 World distribution of salt-marshes and mangrove-swamps. (After Chapman 1977.)

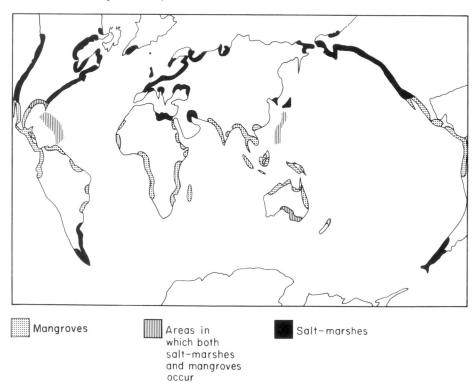

[shading pattern] Mangroves   [shading pattern] Areas in which both salt-marshes and mangroves occur   [shading pattern] Salt-marshes

on the one hand, and sea-grass meadows on the other, is that the first two are intertidal and rarely extend below mean tide level, whilst the sea-grasses extend well below low water to depths of some 15 m, and 30 m in clear water (a few sites appear to support living sea-grasses at depths of 60 m and even 90 m in one species). All three, however, are dominated by tracheophyte plants which can achieve productivities of up to 8500 g (ash-free dry weight) m^{-2} yearly, and all three floras comprise a suite of phylogenetically unrelated species which have convergently adapted to life in their respective habitats. The terms 'mangrove' and 'sea-grass' therefore have no phylogenetic validity: some twenty tracheophyte families (in twelve orders) have produced mangroves; and six families of monocotyledons are represented by grass-like plants which can survive submerged in sea-water (none, however, is a true grass).

4.1   Salt-marsh and mangrove-swamp

Although salt-marshes are dominated by grasses, herbs and dwarf shrubs whilst mangroves are large shrubs or trees (Fig. 1.19), they share many characteristics and hence can be considered together. Both facilitate sedimentation by reducing local water velocities and by retaining many of the particles deposited. Development from a muddy or sandy intertidal zone is initiated by germination of the most salt or submersion tolerant of the component species. These begin the process of trapping sediment (if not already begun by the benthic microflora), thereby raising the level of the shore and decreasing the time of tidal submergence. Other less tolerant, but often competitively stronger species can then establish themselves, similarly to be ousted later by yet further species as shore levels continue to rise. Thus a succession of species results with passage of time which may, partly, be reflected by a zonation of species in space as the marsh or swamp progressively colonizes the upper part of a shore. Cycles of erosion and accretion are common although the causes of the different phases are largely unknown. Colonization or re-colonization of new areas is effected by vegetative spread or by water-borne seeds or fragments. The physiography of both systems is dominated by a network of creeks which ramify through the densely vegetated areas (Fig. 4.2) and provide the route through which sea-water floods and ebbs. Many species characteristic of shallow, marine sediments in general live within these creeks, in pools enclosed by the plant cover, on the marsh or swamp surface, and—in mangroves—as an epifauna or epiflora

Fig. 4.2 Aerial photograph of a salt-marsh (Norfolk, UK) showing the dendritically branching creek systems. (Cambridge University Collection, copyright reserved.)

on the extensive mangrove trunks and aerial roots. These species include single-celled, filamentous and mat-forming algae, deposit-feeding invertebrates and a variety of suspension feeders. The aerial shoots and branches, however, are essentially terrestrial habitats and support an essentially terrestrial fauna. The surface of the sediment therefore forms an interface between the marine and terrestrial ecosystems and there crabs and prosobranchs share the environment with insects, spiders and pulmonate snails.

Both vegetational assemblages are also highly productive: published values generally cluster around 3 kg (ash-free dry weight) m^{-2} yearly. In high latitudes light is the limiting factor to production in winter, whilst elsewhere and at other times inorganic nutrient levels probably set the ultimate limit. Nutrient shortage is only relative, however. Since the plants are rooted and fixed whilst sea-water fluxes tidally through the system, nutrients are present in greater abundance than in most other marine habitats. Grazing pressure on the tracheophytes appears insignificant and is confined to the aerial portions of the plants which are consumed by terrestrial insects (e.g. grasshoppers) and vertebrates (e.g. leaf-eating birds and monkeys). The greater proportion of the plant production (excluding any associated algae) is, therefore, available for consumption only after conversion to detritus. From our present standpoint, the important question is how much of the productivity of these coastal marshes and swamps is transported to the adjacent sea.

Early studies (in the 1960s) indicated that some 50% of the net plant production of these areas was exported by tidal flushing and, for example, a *Spartina* marsh in Georgia, USA, was calculated to export 14 kg ha^{-1} of organic matter per tidal flushing during periods of spring tide and 2.5 kg ha^{-1} during neaps. The water ebbing from this salt-marsh contained between two and three times the quantity of organic matter present in the flooding water. A series of comparable findings in similar regions suggested an export rate of 0.9–1.3 g C m^{-2} daily for salt-marshes and 0.5–2.4 g C m^{-2} daily for mangrove-swamps. If this was generally true, then the marshes and swamps of the coastal fringe would supply some 100 g C per year to each square metre of the continental shelf; a quantity equivalent to the input from the neritic seas. More recently, however, it has become appreciated that not all marshes behave in this fashion (Nixon 1980).

Intertidal marshes may be arranged in a continuum on the basis of their geomorphological and hydrographic setting. At one extreme are those which are open to the sea, are subject to offshore winds and are dominated by patterns of water movement tending to move floating material away from the coast; at the other are those enclosed within land-locked bays, subject to onshore winds and dominated by water movements tending to move flotsam landwards and to strand material along high-level drift lines. The former, open type, including the classic *Spartina* marshes of the coasts of Georgia and South Carolina (Table 4.1), produce an export of salt-marsh detritus, whilst the latter, sheltered type, exemplified by several of those around the southern shores of the North Sea, may actually import fixed organic matter from elsewhere. These two types are the extremes of a

Table 4.1 A general budget for the salt-marshes of the southern and southeastern states of the USA. (After Weigert 1979.)

Net *Spartina* production	1573 g c m^{-2} yearly
Net algal production	180
Total	1653
Metabolized on the marsh	773 g C m^{-2} yearly
Exported by tidal action	880
Total	1653

Table 4.2 A nitrogen budget for a Massachusetts salt-marsh. (After Teal 1980.)

	Input (%)	Output (%)	Balance kg × ha^{-1}
Rain	0.5		3.9
Run-off	22.5		152.2
Nitrogen fixation:			
algal	1.0		6.3
bacterial	9.5		65.2
Tidal exchange:			
NO_3 and NH_4	21.0	30.0	−78.6
particulate	45.0	49.0	−65.2
Denitrification		18.0	−134.8
Sedimentation		3.0	−24.1
NH_4 volatization		0.5	−0.4
Shellfish harvest		0.5	−0.4
Gull faeces	0.5		0.4
			Overall −75.5

continuum, however, and several marshes may be more nearly in balance (Table 4.2). Even this is an over-simplification in that the import–export balance of temperate marshes, at least, may vary within a single year, importing some materials at some times and exporting them at others. This is well shown by a salt-marsh on Long Island Sound, New York (Woodwell *et al.* 1979), in which fluxes of nitrogen, phosphorus, and dissolved and particulate organic carbon included both periods of import and export at different times of the year (Fig. 4.3). Overall, 53 g C m^{-2} yearly

Fig. 4.3 Net tidal exchanges of phosphorus, nitrogen and dissolved and particulate organic carbon between the Flax Pond salt-marsh and the adjacent Long Island Sound, New York. Negative values are losses to the sea and positive values gains by the marsh. (After Woodwell *et al.* 1979.)

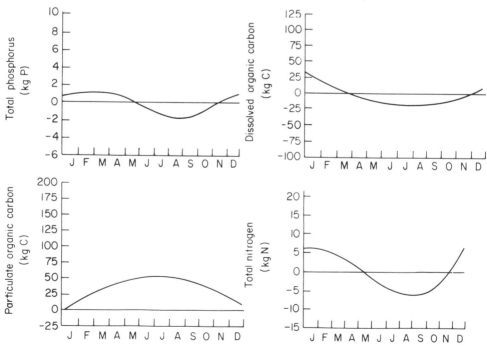

were imported, and 1 g N m^{-2} yearly and 0.25 g P m^{-2} yearly were lost. Large fragments of organic debris showed a net loss, but there was a net gain in finely particulate organic carbon. Table 4.3 summarizes annual fluxes in organic carbon and nitrogen in salt-marshes on the Atlantic seaboard of the USA.

Table 4.3   Annual fluxes of organic carbon and nitrogen between the salt-marshes and coastal waters on the eastern and south-eastern coasts of the USA. Negative values are net exports; positive values are net imports. (From data in Nixon 1980.)

	Carbon (g C m^{-2} yearly)		Nitrogen (g C m^{-2} yearly)	
	Dissolved	Particulate	Dissolved	Particulate
Great Sippewissett, Massachusetts		− 76	− 9.8	− 6.7
Flax Pond, New York	− 8.4	61		
Canary Creek, Delaware	− 38	− 62	− 0.9	− 2.9
Gott's Marsh, Maryland		− 7.3	− 2.1	− 0.3
Ware Creek, Virginia	− 80	− 35	− 2.3	0
Carter Creek, Maryland	− 25	− 116	− 9.2	4.6
Dill Creek, South Carolina		− 303		
Barataria Bay, Louisiana	− 140	− 25		

Clearly, one cannot generalize the role of these intertidal tracheophyte communities: some do indeed produce a large export of material to the coastal sea and thereby partially fuel the food webs of shelf seas, but others are net sinks for organic matter not exporters. They are all sinks in another sense. Grazing on the coarse tracheophytes being insignificant, food webs on the marshes and swamps themselves are in part based on the products of plant decay. These accumulate on the sediment surface and if not remineralized before being buried by sedimentation, become incorporated into the sediment itself. Thus in Woodwell *et al.*'s study, 37% of the net salt-marsh production was buried beneath the surface and Paviour Smith suggested that the ratio of animal to plant to dead organic matter biomass which formed a New Zealand salt-marsh was in the order of 1 : 30 : 680.

The fact that large particles of debris may be distributed differentially to finely particulate material (as in the Long Island marsh) introduces the question of the precise nature of the food available on, and as an export from marshes and swamps. The larger angiosperms produce much litter in the form of dead leaves, etc., and most work has concentrated on this component of the organic input into marine food webs. Such litter decays relatively slowly (Fig. 4.4.) and is soon leached of those soluble

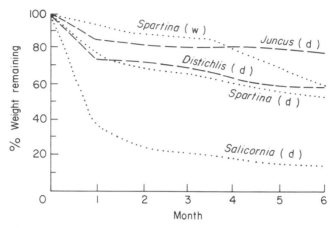

Fig. 4.4 Decay rates of salt-marsh plants in Floridan waters, USA. d = dried; w = fresh materials. (After Wood *et al.* 1969.)

organic compounds remaining in it (many angiosperms translocate much of the soluble material into their root systems before leaf fall). The litter is then colonized by bacteria and fungi in much the same manner as outlined on pp. 78–9 (see Table 4.4). Large particles may often be moved, either to the adjacent sea or onto tidal strand lines, before complete decomposition has occurred. Smaller particles created by mechanical or biological breakdown are available to be consumed by deposit feeders, however, although it seems unlikely that these are the preferred food source on the marshes themselves. The 'detritus-feeding' gastropods and crabs of vegetated intertidal zones probably subsist largely on the abundant (but somewhat neglected) algae which are invariably associated with the larger angiosperms, growing

Table 4.4 Microbial degradation of mangrove leaves in a Florida estuary. (After Cundell *et al.* 1979.)

1–14 days	Rapid leaching of sugars and tannins.
14–28 days	Depletion of leachable sugars; continued leaching of tannins; initial colonization by bacteria.
28–49 days	Depletion of leachable tannins; colonization of outer leaf surface by bacteria, fungi and protists.
49–70 days	Erosion of leaf surface; rich microbial community of cellulytic bacteria and fungi; many protist and meiofaunal organisms.
> 70 days	Fragmentation of leaves.

over their surfaces and on the sediment between the plants. The biomass of all these algae may exceed 25% of that of the salt-marsh tracheophytes (but a much lower percentage of total mangrove-swamps biomass), and productivities of the microalgal benthos can reach—and probably exceed—30% of that of their larger associates. Recent evidence suggests that littorinid, hydrobiid and nassariid gastropods and a variety of crustaceans graze this microalgal productivity as fast as it is produced. The larger plants may therefore contribute relatively little to *in situ* secondary production, less even than the phytoplankton stranded on the sediment and amongst the vegetation by the tide.

4.2   Sea-grasses

Sea-grasses are more thoroughly marine and only a few species extend into the intertidal zone. Their productivity and its fate are therefore entirely marine phenomena and it is less realistic to discuss them in terms of exports and imports. Nevertheless, dead sea-grass leaves can be transported over huge distances. They have been recorded at depths of nearly 8000 m; sea-grass remains were identified in nearly every one of a series of 5300 photographs taken of the sea bed at an average depth of 4000 m; and after hurricanes, mats of leaves up to 50 m across have been reported from the Florida Current (see p. 112). Yet in life they are confined to the littoral zone. Long-distance transport of leaves clearly depends on hydrographic conditions, but it also varies with the properties of the leaf itself. Some species (e.g. *Thalassia*) sink after being detached, whilst others (e.g. *Syringodium*) float, at least for some time (Fig. 4.5). Thus in the sea-grass meadows of the Virgin Islands, for example, *Syringodium* leaves account for 92.5% of the loss of sea-grass material

Fig. 4.5 The effects of grazing by parrotfish (*Sparisoma*) on two sea-grasses: *Thalassia* and *Syringodium*. (After Zieman *et al.* 1979.)

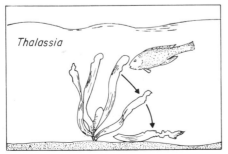 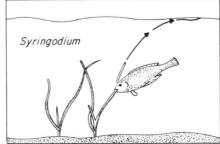

$(15 \times 10^4$ g or 0.2–0.4 g m^2 daily) even though *Thalassia* comprises 94% of the total sea-grass biomass and 90.5% of the productivity.

The term meadows aptly describes the appearance of stands of sea-grasses. Up to 15000 shoots of the smaller species, with several leaves per shoot, may arise from each square metre of subtidal sand or mud flats, the leaves often being strap-shaped. Biomass and productivity vary latitudinally, being greatest in tropical waters. In temperate areas, mean biomass probably lies near to 500 g dry weight m^{-2} (with a maximum recorded value of over 5 kg m^{-2}) and productivity is of the order of 2 g C m^{-2} daily during the growing season (up to a maximum of less than 10 g C m^{-2} daily). Comparable figures for tropical sea-grasses are: biomass averaging over 800 g dry weight m^{-2} (maximum 8 kg m^{-2}) and productivity averaging some 5 g C m^{-2} daily, with a maximum of nearly 20 g C m^{-2} daily. Similar to the macrophytes considered in section 4.1, few animals appear directly to use sea-grass production as food; some fish, turtles, sirenians and a few sea-urchins with ruminant-like cultures of cellulose-splitting bacteria in their guts are notable exceptions. Once again, however, most consumers are dependent on the decomposing grasses, and bacteria are critical in rendering the plant material digestible. Experiments have shown that aged sea-grass detritus is assimilated to a much greater degree than 'artificial detritus' prepared by macerating living sea-grass leaves (thereby permitting access to the cell contents). Perhaps 5% of sea-grass production is consumed directly: a higher value than that of the coarser salt-marsh and mangrove-swamp plants with their more extensive supporting tissues.

Sea-grasses decay slowly (otherwise they would not occur in identifiable form in the deep sea) but nevertheless more quickly than intertidal tracheophyte material (Fig. 4.6). In the tropics, *in situ Thalassia* leaves may loose 10–20% of their initial dry weight per week (varying with the extent to which they are broken down mechanically by water movement) and hence they would be completely decomposed in less than one year. Decomposition is slower in higher latitudes and there eel-grass (*Zostera*) leaves may only loose 20% of their original weight in 100 days and much eel-grass material will be remineralized in and below the redox discontinuity layer. Because sea-grasses are normally submerged, the products of their decay can enter the planktonic system to a larger extent than can, for example, leaves of salt-marsh plants. At times of leaf die-back and decomposition, heterotrophic protists abound in the water column over the meadows and take up both the dissolved organics leached from the sea-grasses and the rapidly multiplying bacteria. Sea-grass detritus is also rich in micro-organisms. One gram (dry weight) has been calculated to support, on average, 10^9–10^{10} bacteria, $5 \times 10^7 - 10^8$ hetero-trophic flagellates and $10^4 - 10^5$ ciliates, yielding a total biomass of some 9 mg of monerans and protists to deposit feeders.

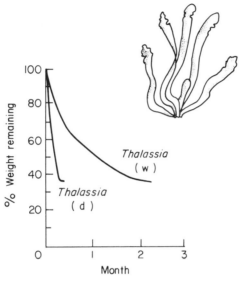

Fig. 4.6 Decay rates of the sea-grass *Thalassia* in Floridan waters, USA. d = dried; w = fresh materials. (After Wood *et al.* 1969.) Compare with Fig. 4.4.

Living sea-grass leaves also provide an attachment site for nu-merous types of epiphytic algae (both macroscopic and unicell-ular), and other algae occur between the sea-grass shoots and in the surface layers of the sediment. As in the other tracheophyte communities, it is these softer, more digestible algae which sup-port the abundant grazers associated with the meadows. In rela-tively open stands, the benthic algae may account for 70% of the total primary production, but thick carpets of leaves reduce the light available to the algal understory (Fig. 4.7) and in such re-gions their productivity is less than on open expanses of sedi-ment. Estimates of epiphytic productivity are relatively scarce but biomasses of the same order as those of the leaves to which they are attached are known. An average contribution of over 15% of the total carbon fixation is not unlikely.

Sea-grass meadows are therefore the most productive of shal-low, sedimentary environments and not only does their primary production support a rich, resident fauna, for which reason they are frequently used as nursery areas by nektonic species (p. 205),

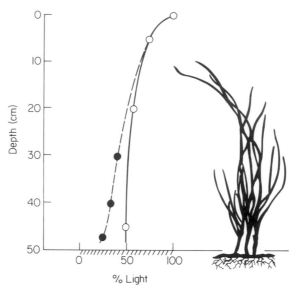

Fig. 4.7 Light penetration through shallow waters in the absence of eel-grass, *Zostera* (solid line), and when eel-grass causes self-shading (dashed line). (After Short 1980.)

but it also consistently enters into the food webs of areas far removed from the coastal zone. A question mark does still hang over the importance of exports from all three of the stands of vegetation considered in this chapter, however, and any error is likely to be in the direction of underestimation. Most measurements of loss rates of fixed organic matter are carried out under reasonably clement weather conditions (for understandable reasons). Yet it is likely that freak weather (hurricanes, etc.) can remove more tracheophyte material in one hour than is accomplished during one or more years of normal wind and wave conditions, although for practical reasons it is impossible adequately to sample these irregular storm losses. One such measurement has indicated that, during severe storms, losses of debris from salt-marshes can be at the rate of over 11 kg ha^{-1} h^{-1}. Observations like this and that after the hurricane in the Florida Current (p. 109) cannot do other than suggest that estimates which do not take storm losses into account must seriously underemphasize the transport of material from the coast to the open sea.

5 Rocky Shores and Kelp Forests

Rocky shores and kelp forests are treated sequentially in this chapter because, in cool, temperate regions, the one type of ecosystem is often an extension of the other. Macroscopic algae are the main primary producers, fucoids intertidally and kelps subtidally. Underlying mechanisms of zonation and community organization are basically similar in both ecosystems, so that examples from each can be used in a complementary fashion. The common thread is the predominance of sedentary organisms, a feature also shared by coral reefs that replace kelp forests in suitable tropical environments; indeed many of the principles of community organization encountered in this chapter will appear again in slightly different forms in Chapter 6.

5.1 Rocky Shores

5.1.1 Colonization of the marine–terrestrial gradient

Most students, if asked to describe the biological features of a rocky shore, would mention zonation. This spatial sequence of organisms up and down the shore is universal, involving much the same sort of organisms anywhere in the world. Why should this be so? The shore is the transition between marine and terrestrial environments, but because of water movement associated with tides, waves, and spray, the transition is gradual, creating a habitat in which neither fully marine nor fully terrestrial organisms can prosper. Emersion is stressful to marine organisms, just as immersion is stressful to terrestrial ones. Organisms able to withstand different degrees of stress replace each other along the marine–terrestrial gradient, and this fundamental biological phenomenon applies on any rocky shore from the tropics to the poles. Alternating inundation and exposure to the air are the environmental features of overriding importance determining the kinds of organisms living on rocky shores, and that is why rocky-intertidal organisms are rather similar world-wide, despite gross dissimilarities in climate.

One might expect that any transitional gradient between two dissimilar environments would be colonized more or less equally by organisms from either source, yet in rocky intertidal habitats the great majority of organisms are of marine origin. The intertidal zone, however, does not embrace the entire gradient between marine and terrestrial environments. The effects of wave splash and spray extend beyond High Spring Tide Level, and organisms in this spray zone are predominantly of terrestrial origin.

5.1.2 *Patterns of zonation*

In spite of its ecological importance as the 'other half' of the marine–terrestrial transitional gradient, the spray zone has been studied relatively little, so that most of what is known about life on rocky shores pertains to the intertidal zone. People familiar with temperate shores will be impressed by the dark festoons of fucoid algae covering much of the intertidal rock surface in moderately sheltered localities. The fucoids illustrate very well the phenomenon of zonation (see Fig. 5.4) that is characteristic of intertidal sedentary organisms. Mobile animals also tend to be zoned but their limits are not so clearly demarcated as are those of sedentary animals. This is to be expected since mobile animals can seek refuge in crevices, under weed, or beneath stones at low tide and these protective microhabitats tend to be spread throughout the intertidal zone.

Zonal variation between shores

Zonation patterns vary among shores. The order in which species are zoned remains the same, very rarely do species transpose positions down the shore, but the widths of the zones and their degree of overlap vary, while some species disappear altogether and may be replaced by others.

Both physical and biological factors may be responsible for this variation. The biological factors, competition and predation, differ in importance from place to place, often interacting in subtle ways with physical environmental conditions and are dealt with in section 5.1.5. Zonal width must ultimately depend on the extent of suitable microclimatic conditions. For sedentary organisms, several physical factors may influence the extent of suitable microclimates. First, the slope of the shore will determine the amount of suitable rock surface; the flatter the shore the greater the width of substratum lying between a given interval of tidal height. Second, tidal amplitude will be of great importance. Most localities around Britain have a rise and fall of 3–6 m on spring tides, but the range may be as much as 9–12 m in the Bristol Channel and as little as 0.6 m off south-west Argyll. The greatest tidal range in the world is in the Bay of Fundy, Nova Scotia, where the spring tidal range may exceed 15 m. In central regions of the Red Sea there is virtually no tidal rise and fall, but because wave action wets the rock surface, intertidal organisms occur, but their zones are narrower than at localities experiencing larger tides (Fig. 5.1). Third, the average amount of

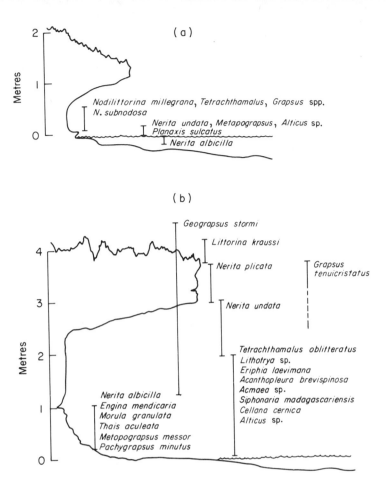

Fig. 5.1 Increasing tidal amplitude vertically extends the intertidal habitat, allowing more species to colonize the shore. (a) Mid way along the Red Sea where there is no appreciable tidal oscillation; (b) Aldabra atoll, Indian Ocean, where there is a spring-tidal amplitude of 3 m. (After Hughes 1977.)

wave action will influence the degree of wetting a substratum receives at a given tidal height; the greater the exposure to wave action the further the zonal boundaries are pushed upshore (Fig. 5.2). Fourth, the aspect of the shore will affect the amount of drying out at low tide; on shaded slopes surfaces will remain damp for longer than on sunny slopes, so that other things being equal upper zonal limits should be somewhat higher.

These factors will also affect the zonal width of mobile animals in a similar way, but an additional variable for them is the availability of suitable refuges that are used during low tide. *Littorina neritoides*, for example, is sometimes confined to rock crevices above Mean High Tide Level, but on shores with an abundance of empty barnacle shells, which can be used as an alternative to rock crevices, *L. neritoides* may extend down to Low Spring Tide Level.

115

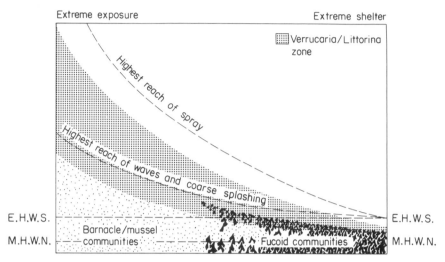

Extreme exposure   Extreme shelter

Verrucaria/Littorina zone

Highest reach of spray

Highest reach of waves and coarse splashing

E.H.W.S.   E.H.W.S.
Barnacle/mussel
M.H.W.N. communities   Fucoid communities   M.H.W.N.

Fig. 5.2 Increasing exposure to wave action elevates zonal boundaries and increases the vertical extent of the splash zone. E.H.W.S. = Extreme High Water Spring Tide; M.H.W.N. = Mean High Water Neap Tide. (After Lewis 1964.)

Superimposed on changing zonal boundaries between shores is the replacement of organisms occupying equivalent tidal heights. In Britain this amounts to the predominance of fucoid algae on more sheltered shores and their replacement by barnacles and mussels on exposed shores (Fig. 5.2). This is ultimately the result of wave action, which on exposed shores tears algal fronds from the substratum, leaving space available for barnacles and mussels.

Geographical patterns of zonation

On a world-wide basis, the characteristics of rocky-intertidal faunas and floras are influenced by gross changes in climate in addition to the factors already considered.

In polar regions, ice scrapes the rocks for much of the year and organisms survive only in crevices and hollows, or by migrating to the sublittoral zone. Deep freezing during winter and dilution during the summer thaw create a harsh environment. Very little macroscopic life occurs in the spray zone and is confined to summer flushes of annual algae in temporary pools. The mid-littoral zone becomes colonized by annual filamentous and membranous algae that are cropped by a few herbivores including amphipods, littorinids and limpets. A much richer algal flora occupies the sublittoral zone, some species reaching very large sizes, and among this lush vegetation are representatives of most of the major animal groups present in temperate latitudes.

In tropical regions, high light intensities, high temperatures and desiccation at low tide are too severe for most macroscopic algae to survive high in the intertidal zone. In contrast to the festooned rocks of temperate shores, tropical intertidal rocks seem rather desolate at first glance, and it is only below Mean Low Tide Level that life takes on the richness one expects in the tropics (Chapter 6).

The tropical spray zone lacks the tufted lichens characteristic of temperate regions, but is colonized by encrusting lichens and blue-green algae. Littorinid gastropods graze the flora when dampened by spray or by rain. At night, small, terrestrial hermit crabs scavenge over the area and carnivorous grapsid crabs come up from the intertidal zone to search for prey.

The tropical mid-intertidal zone is colonized by blue-green and filamentous green algae. So close is the growth that often only the greenish colour of the rock surface indicates the presence of plant life. Nevertheless, the productivity of the intertidal algae is evidently very high, because it supports a substantial guild of grazers. Many of these animals are similar to those in other latitudes, for example limpets, chitons, winkle-like gastropods, isopods, and amphipods, but in addition there are grazers not present in cool climates. Among these are the grapsid 'sally lightfoot' crabs that have spoon-shaped claws with which they scrape algae off the rock surface during low tide. Crabs that forage out of the water, like grapsids and terrestrial hermit crabs, peter out in warm, temperate regions. Presumably at higher latitudes cool air temperatures would lower their metabolic rate too much, causing the crabs to lose their vital agility (see Chapter 10). Further details of British shores are to be found in Lewis (1964) and of world-wide shores in Stevenson and Stevenson (1972).

5.1.3 Determinants of zonation

Why are rocky intertidal organisms zoned? There are two levels of approach to this question, one concerning ultimate factors that prevent both the random mingling of species and the monopoly of the intertidal habitat by a few 'super organisms', and another concerning proximate factors that impose spatial limits to the zones.

Ultimate factors

Two ultimate factors are responsible for the general phenomenon of zonation: competition for resources and the restricted potential of any species to perform optimally under different environmental conditions. Competition among species for a common limiting resource is an inevitable process if populations overlap and are not kept at very low densities by grazing, predation or physical factors. Prolonged competition will cause populations of the better competitors to expand and those of the poorer competitors to dwindle to local extinction. Competitive ability depends on various aspects of performance, for example growth rate, growth form, reproductive rate, aggressiveness, feeding efficiency, and to excel in any such field requires a degree of specialization. Specialization in one field is usually achieved at the expense of poorer performance in another (e.g. fucoids, p. 120, barnacles, p. 124, limpets, p. 125). High performance at one position along an environmental gradient will therefore be accompanied by poorer performance elsewhere. Different species will perform best at different positions, each with the potential, to exclude the others from its optimal location. Thus we find that species replace one another, i.e. are zoned, along any pronounced environmental gradient, whether it be the marine–terrestrial transition, or any other kind of transition, such as one of salinity along an estuary, depth on a coral reef (p. 165), altitude on a mountainside or aridity at the edge of a desert. Zonation is a particular manifestation of resource-partitioning among potentially competing species. Linear spatial replacements (zones) reflect the strongly linear nature of the environmental gradient, and in the absence of pronounced gradients, resource-partitioning results in more complicated spatial patterns among species (e.g. tropical predatory gastropods, p. 181), but the ultimate factors of competition and specialization remain the same.

Proximate factors

Proximate factors fixing the limits to zones could, in principle, be: (a) behavioural, where the organism 'chooses' to settle or remain within its zone of optimal performance; (b) physiological, where the organism is unable to survive in the environmental conditions outside its zone; or (c) ecological, where the organism's potential distribution is curtailed by competition or predation. In practice, remarkably little is known about the proximate factors affecting zonation in many of even the commonest rocky shore organisms. It is so easy to make the unjustifiable assumption that because organisms from higher levels on the shore have greater physiological tolerances to desiccation or extremes of temperature and salinity when subjected to these stresses in the laboratory, then the same factors must limit distribution on the shore. For many years, people tended to be satisfied with demonstrating the almost inevitable correlation between physiological tolerance to stress and zonal height on the shore (e.g. fucoids, p. 119, barnacles, p. 124, limpets, p. 125). Yet as a rule, environmental conditions measured on the shore are far less extreme than the limits of tolerance measured in the laboratory. Not considered, was the alternative possibility that the correlation between physiological tolerance to laboratory-induced stress and zonal height on the shore reflects the adjustment of physiological mechanisms so that they work optimally under the appropriate natural environmental conditions. This interpretation is congruent with the concept of ultimate factors determining zonation: optimal physiological performance at a particular tidal level being sharpened on an evolutionary time-scale by interspecific competition, but costing poorer performance at other levels.

Among the relatively few species for which the proximate factors of zonation have been identified, it appears that physiological factors often fix upper zonal limits, especially of high-shore species, but that ecological and behavioural factors tend to fix lower zonal boundaries.

5.1.3.1 Zonation of macroalgae

Optimal performances at the appropriate heights on the shore, reflecting the ultimate factors of zonation, have been demonstrated for a series of Californian intertidal algae. The capacity of these algae to sustain photosynthesis in air varies according to zonation. Species from the lower shore photosynthesize less well in air than under water. Species from the mid to upper shore reach maximum photosynthesis after some degree of drying, whereupon the photosynthetic rate in air can be as much as six times that under water at the same illumination and temperature, although the physiological mechanism for this remains unknown. Further drying retards photosynthesis, and the time taken for desiccation to reduce the photosynthetic rate to half its maximum value increases successively in algal species from higher levels on the shore (Fig. 5.3). These differences are due to lower desiccation rates of the higher shore algae rather than an increased tolerance of desiccation itself. Desiccation rates are low enough under normal conditions that these algae are capable of sustaining high rates of photosynthesis throughout tidal emersion, when most of their carbon is probably fixed.

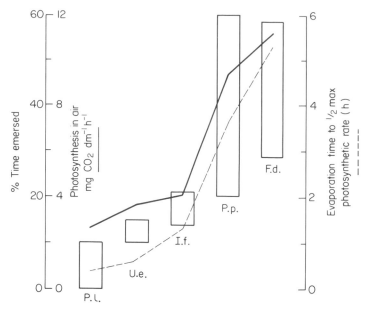

Fig. 5.3 Californian algae from an increasing zonal height on the shore exhibit an increased capacity to maintain photosynthesis during tidal emersion. P.l. = *Prionitis lanceolata*, U.e. = *Ulva expansa*, I.f. = *Iridaea flaccida*, P.p. = *Porphyra perforata*, F.d. = *Fucus distichus*. (After Johnson *et al.* 1974.)

Zonation of Scottish algae

Five fucoid algae occupy overlapping but well-defined zones on a sheltered rocky shore on the Isle of Cumbrae, Scotland (Fig. 5.4). As with the Californian algae, the ability to tolerate exposure to desiccating conditions, and then maintain normal growth, increases progressively in species from low to high shore levels. Among naturally situated plants, damage is sometimes caused during the spring and summer when neap tides leave the upper shore continually emersed for several days of calm, warm, dry weather, leading to the death of *Ascophyllum nodosum*, *Fucus spiralis* and *Pelvetia canaliculata* at the upper edges of their zones. Prolonged exposure by neap tides to winter frost and rain, or even hard freezing, has no ill effects. Schonbeck and Norton (1978, 1980) found that *Fucus spiralis* transplanted into the *Pelvetia* zone grew poorly and eventually died even though not crowded by *Pelvetia* itself. Desiccation is evidently an important proximate factor determining the upper zonal boundaries of *Ascophyllum nodosum*, *Fucus spiralis* and *Pelvetia canaliculata*. Drought damage was not observed among *Fucus vesiculosus* or *Fucus serratus*, and this, together with the less clearly defined

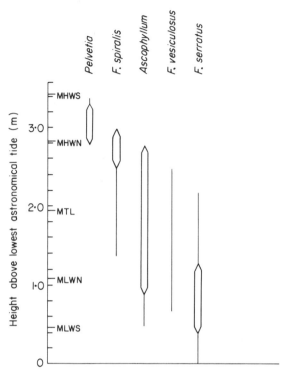

Fig. 5.4 Zonation of fucoids on the Isle of Cumbrae, Scotland. MHWS = Mean High Water Spring Tide; MHWN = Mean High Water Neap Tide; MTL = Mean Tide Level; MLWN = Mean Low Water Neap Tide; MLWS = Mean Low Water Spring Tide. (After Schonbeck & Norton 1978.)

upper zonal limits, suggests that ecological factors may curtail the upshore distribution of these algae below the potential level set by physiological factors.

Pelvetia and *Fucus spiralis* transplanted downshore grew well (Fig 5.5a), proving that the lower zonal boundaries of these algae are not determined by physical environmental conditions.

Factors limiting the lower zonal boundaries of the other fucoids were not investigated experimentally, but the generally increasing growth rates of species from lower shore levels (Fig. 5.5b) suggests that competitive exclusion may be a common mechanism.

Zonation of Australian algae

On sheltered rock platforms in New South Wales, foliose macro-algae such as *Ulva lactuca*, *Colpomenia sinuosa* and *Corallina officinalis* cover most or the substratum at low-shore levels, but

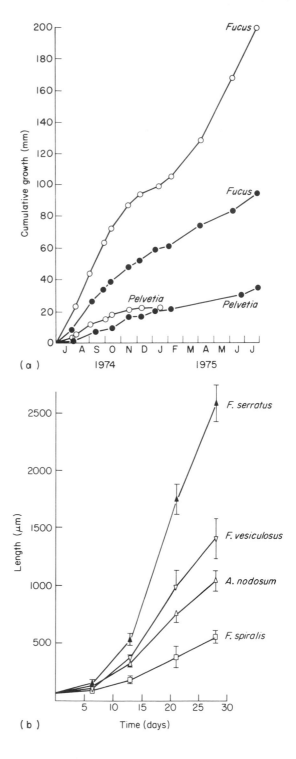

Fig. 5.5 (a) Cumulative linear growth of *Pelvetia canaliculata* and *Fucus spiralis* transplanted into their own zones (solid circle) and onto the middle shore (open circle). (b) Linear growth of fucoids cultured from zygotes in the laboratory. (After Schonbeck & Norton 1980.)

peter out abruptly in the midshore region. Here the rock is extensively covered by the encrusting red alga *Hildenbrandia prototypus* and supports dense populations of grazing gastropods. Harsh physical factors, such as desiccation and high temperatures due to intense insolation at low tide, or intense grazing by the numerous gastropods, could limit the foliose algae to low shore levels. To determine the relative importances of these factors, Underwood (1980) constructed a series of stainless steel mesh fences and cages on the rock platform. The fences excluded grazers but cast virtually no shade, whereas the cages both excluded grazers and cast sufficient shade to reduce the temperature of the covered areas. Roofs were also constructed like cages without the sides, so as to cast shade without excluding grazers. The experiments demonstrated that the grazing gastropods were removing all spores and sporelings of foliose macroalgae in the midshore region, and that even when grazing was prevented, harsh physical factors killed the algae as they grew beyond the sporeling, 'algal turf' stage.

The encrusting alga *Hildenbrandia* can withstand the physical conditions and although gastropods graze over its surface they merely remove spores of other algae.

5.1.3.2 Zonation of animals

Zonation of sedentary animals

Proximate factors determining the zonal boundaries of intertidal algae often also determine those of intertidal animals, but the zonal distribution of animals may be influenced by two additional features: mobility and food. After settlement and metamorphosis, many sedentary animals such as sponges, hydroids, tunicates, tubicolous polychaetes, vermetid gastropods, oysters and barnacles, are immobile, although barnacles can be pushed along to some extent by pressure from growing neighbours. Other sedentary animals such as sea anemones and mussels are capable of very slow movement and therefore have a limited capacity to adjust their position according to circumstances, while non-sedentary animals are potentially capable of moving throughout the intertidal zone within one or a few tidal cycles. Increasing mobility may be expected to reduce the effects of physical environmental stress on zonation, which may be influenced more by the distribution of protective microhabitats or food supply. Of course, the latter are also important resources for sedentary animals: the distribution of protective microhabitats may influence larval settlement, and the daily food supply of sedentary animals, all of which are 'particle' feeders, will decrease with the duration of tidal immersion upshore.

Mussels

Whether time available for feeding ever limits the intertidal distribution of filter-feeders is unknown, but its effect on the growth of mussels is apparent. *Mytilus edulis* from high-shore levels may live for many years, but grow very slowly and never reach a large size. When transplanted to low-shore levels, the same mussels put on a spurt of growth and achieve large sizes characteristic of low-shore mussels. Large *Mytilus edulis* transplanted from low- to high-shore levels die, being unable to acquire sufficient food to maintain their large biomass. Desiccation stress, rather than limited opportunities for feeding, probably fixes the upper zonal limit of *Mytilus edulis* on most shores, however. Hundreds of thousands of *M. edulis* at the upper edge of the mussel zone on parts of the Washington coastline have been found dead and gaping during extremely hot summer days. This summer mortality causes the upper edge of the mussel zone to move several cm downshore as the dead shells are gradually washed away. During late winter to early spring, mussels move up from the main population to occupy the newly vacated space, a behaviour presumably driven by intraspecific competition. The lower boundary of the *M. edulis* zone on this shore is set by starfish predation. In subtidal regions *M. edulis* survives in refuges among patches of filamentous algae, hydroids, bryozoans, or in kelp holdfasts where starfish cannot reach them.

Barnacles

Two common barnacles on British shores, *Balanus balanoides* and *Chthamalus montagui* (previously referred to as *stellatus*), are limited by physical factors in their upward distribution and by biological factors in their downward distribution on the shore. Cyprids of both species settle on suitable substrata throughout most of the intertidal zone down into the sublittoral zone. *Balanus* cyprids can cement down their antennules on the rock surface in about 20 min and are therefore able to settle at high-shore levels where tidal immersion is brief. However, the highest settling cyprids are killed by desiccation before the next tidal immersion. Resistance to desiccation increases as the barnacles grow, but even adults may be killed by desiccation during calm spring or summer weather coinciding with neap tides (cf. *Pelvetia* p. 121). As a result, the upper zonal boundary of adult *Balanus* is considerably below the limit of larval settlement. Cyprids of *Chthamalus montagui* apparently can settle out on surface films left by wave surge, enabling them to attach at higher shore levels than *Balanus*. Resistance to desiccation at all growth stages is greater in *Chthamalus* so that the adults extend higher on the shore than *Balanus*. By continually removing *Balanus* from a strip of rock-face on the Isle of Cumbrae, Connell (1972) showed that *Chthamalus* was able to settle and grow down through the *Balanus* zone into the sublittoral zone. The lower limit to the natural distribution of *Chthamalus* therefore, is set by competition with *Balanus*. *Chthamalus* normally lives in a competitive refuge from *Balanus* high up on the shore, and is able to do so because of a high temperature tolerance, reduced activity when the tide recedes, a tightly fitting operculum and a non-porous shell. These attributes are gained at the expense of a lower potential growth rate. *Balanus* has a looser, more porous shell that is evidently quicker to produce. On the Scottish shore, the lower zonal limit of *Balanus* is set by predation, as Connell showed by enclosing the barnacles in stainless steel cages that excluded the predatory dogwhelk, *Thais lapillus*.

Zonation of mobile animals

The mobility of non-sedentary animals makes the study of their intertidal distribution less straightforward. Limpets represent a 'half-way stage', sharing some of the hazards of a sedentary life and some of the benefits of mobility. Zonation of limpets has been examined extensively in Britain, South Africa, New Zealand and Australia, but Wolcott's (1973) study of Californian limpets serves well for illustration. Three species live on the bare rocks in the splash zone. *Acmaea scabra* occupies mainly horizontal surfaces fully exposed to the sun. It 'homes' to a scar on the rock surface and the shell margins grow to form a precise fit to the scar, thereby reducing desiccation. *Acmaea digitalis* occupies primarily vertical or overhanging surfaces shaded somewhat from the sun. *Acmaea persona* retires during the day into dark crannies or beneath boulders, emerging to feed only at night. When not feeding during the day, *A. digitalis* and *A. persona* secrete a mucous sheet between the shell margin and the rock surface. This acts as a diffusion barrier and reduces water loss.

All these limpets respond to desiccation by ceasing movement. *A. scabra* returns to its scar and the others clamp down on the rock surface as it dries out, so that just like sedentary animals, they must endure any stressful conditions until the tide returns. Rock-surface temperatures were never observed to exceed the thermal tolerances of the limpets (which increase in species from higher shore levels, suggesting the adjustment of physiological processes for optimal performance under the appropriate conditions (see p. 119)) and desiccation seems to be the main hazard.

Death ensues through the concentration of the body fluids to lethal levels. The desiccation tolerances of the three splash-zone limpets, amounting to about 80% total water loss, are amongst the highest recorded in any animals and involve physiological adaptation to high electrolyte concentrations at the cellular level. Nevertheless, desiccation sometimes takes its toll by concentrating the body fluids to lethal levels. Hundreds of *A. scabra* and *A. digitalis* up to 11 years old were killed in August 1971 after the sea had been exceptionally calm, leaving the rocks dry for over a week where they would normally be wetted by spray.

Lower boundaries to zones and partial segregation within areas of zonal overlap are probably determined by competition. For example, *Acmaea digitalis*, owing to its ability to secrete a mucous diffusion barrier that can be made to fit any irregular surface, does not need to forage near to a close-fitting 'home site' as does *A. scabra*. *A digitalis* is able to outcompete *A. scabra* throughout most of the splash zone, except in the most exposed, sun-baked places where *A. scabra* has a competitive refuge. *A. persona* feeds only at night, avoiding contact with *A. scabra* and *A. digitalis*.

5.1.4 Population ecology

Sedentary organisms

Space on the substratum is an essential resource for sedentary organisms and may be provided by the rock surface (primary space) or by biological surfaces (secondary space) such as those of algal fronds or mussel shells. Space is non-renewable in the sense that once occupied, no more is forthcoming. This contrasts with renewable resources such as food, which can be replaced as it is eaten by growth of the food organisms. Primary space is fixed in extent, imposing a limit to the number of sedentary individuals that can occupy the rock surface of a given area of shore. Secondary space is not fixed, but may wax and wane according to the population dynamics and individual growth of the organisms providing the surface (see *Fucus serratus*, p. 137).

The availability of primary or secondary space is one of the most important population-controlling factors among sedentary organisms. As space becomes used up, individuals within a population are thrown into competition among themselves (intraspecific competition) and some will compete with other species (interspecific competition). The restriction of space available to an individual may be detrimental in two possible ways: physical and biological. Physical constraints may involve reduced area for attachment, deformed growth, undercutting and smothering or crushing due to the encroachment of adjacent competitors (see *Chthamalus montagui*, p. 124). Biological constraints may involve reduced accessibility to light (plants) or water-born particles or algal films (animals), thereby reducing growth and reproductive output. Important consequences of interspecific competition for space among sedentary organisms are illustrated in sections 5.1.5 and 5.2.4 dealing with community structure.

Intraspecific competition has an effect that is particularly noticeable among sedentary animals occupying primary space, and is caused by the way availability of space interacts with food supply. A unit area of substratum allows access to a certain average supply of food. It follows that the unit area of substratum can support only a limited biomass of sedentary animals, equivalent to the maximum sustainable by the food supply. Populations of sedentary organisms have the capacity to 'achieve' the maximum sustainable biomass over a remarkably wide range of densities. This 'regulation of biomass' results from two processes: (1) 'self-thinning' as weaker competitors within the population die while the rest continue to grow (eg. kelp, *Egregia*, p. 145); and (2) adjustment of growth rate and hence body size in a negatively density-dependent manner, so that individuals reach smaller sizes at higher densities. Foresters have long been aware of these processes among stands of trees.

Lacking the ability to move to better conditions, sedentary organisms have highly adaptable growth forms (Chapter 9) that can be modified to suit local conditions. The barnacle *Balanus balanoides* develops a squat, volcano-shaped body when not crowded, but becomes taller and thinner as crowding increases (Fig. 5.6b). When grown at increasing densities, the final size of *Balanus* decreases, but total biomass remains constant once the 'carrying capacity' of the habitat has been filled (Fig. 5.6a). Correspondence of the total final biomass to the carrying capacity is illustrated by rearing barnacles under conditions with different food supplies: final biomass increases as food supply becomes richer (Fig. 5.6a). At high densities, barnacles become stunted, and since their fecundity is directly proportional to body size, crowded individuals are able to produce fewer eggs (Fig. 5.6c).

A similar effect on growth and fecundity is found in the South African limpet *Patella cochlear*. Although mobile, *P. cochlear* feeds within a territory only a few cm in diameter and therefore behaves as a quasi-sedentary animal. The limpet grazes a 'garden' of red algae in a controlled manner, exploiting the 'interest' but not the 'capital' so that a renewable food resource is sustained. As the limpet grows it expands its territory accordingly, but as population density increases, space becomes limiting, restricting the expansion of territories of new recruits. Final body size (correlated with territory size and the sustainable yield of algae) declines at high population densities and although total biomass per unit area is regulated (Fig. 5.6d), individual fecundity declines (Fig. 5.6e).

Conservation of biomass in *Balanus* and *Patella* is not achieved 'for the good of the populations' but is merely a consequence of competition among individuals coupled with their plasticity of

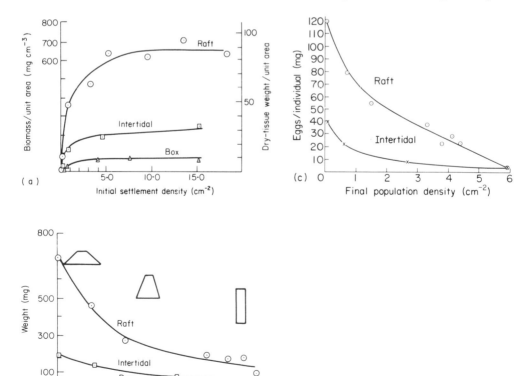

Fig. 5.6

growth. Moreover, the accompanying density-dependent regulation of fecundity (Fig. 5.6c, e) could not normally affect local recruitment, because many larvae are likely to come from populations further afield.

Mobile animals

Potentially limiting resources for mobile, intertidal animals include food and protective microhabitats such as crevices or empty barnacle shells (see *L. neritoides*, p. 116). Quantifying the availability of these resources and their effect on consumer populations proves to be much more difficult than with sedentary organisms using space. Problems arise because the animals move about, making it difficult to assess what conditions they experience during their travels, and because in the case of grazers it may not be clear exactly what they are feeding on. Do they assimilate all detritus, microbes, spores, macroalgal germlings, diatoms and endolithic algae that can be scraped off the rock, or are they more selective? Underwood (1979) circumvented these

Fig. 5.6 (a) Regulation of biomass in the barnacle *Balanus balanoides*. Total biomass per unit area reaches an asymptote when barnacles are grown at increasing densities. The asymptotic biomass increases as the general level of food supply increases. (b) Biomass is regulated because individual barnacles attain smaller sizes at higher densities. Lateral pressure from neighbours causes the body to become taller and more cylindrical. (c) Fecundity is proportional to body size, and so declines in barnacles grown at increasing densities. (d) Regulation of biomass in the limpet *Patella cochlear*. Limpets attain a smaller size when growing at higher densities. (e) Fecundity is proportional to body size and therefore declines among limpets growing at higher densities. (a–c after Crisp 1962; d–e after Branch 1975.)

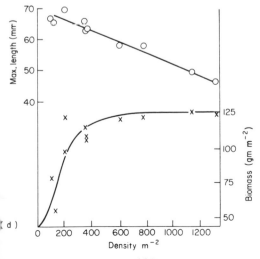

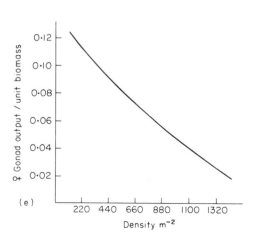

difficulties by quantifying the effect of experimentally controlling crowding on growth and survivorship in several Australian grazing gastropods. When the snail *Nerita atramentosa* was kept in wire-mesh cages fixed to the natural rock-face, the growth rate of juveniles decreased, but their survivorship remained constant, as density was increased. Adults gradually lost weight when kept at high densities and the survivorship declined as density increased. These effects could only have been due to competition for food, although the food supply itself could not be quantified. Thus, in a crowded population, juvenile *Nerita* survive at the expense of adults because their smaller individual biomass can be maintained on less food. Adult mortality is therefore likely to follow periods of heavy recruitment. As with barnacles and limpets, *Nerita* has planktonic larvae that can recruit onto the shore from diverse origins. There can be no feedback, therefore, between local adult density and future levels of recruitment. The unpredictable fluctuations in numbers of recruits, characteristic of planktonic larvae (Chapter 9), will consequently be reflected in the density and size structure of adult populations.

5.1.5 Community structure

Rocky-shore organisms lend themselves particularly well to experimental manipulation. Algae and sedentary animals can be thinned out, whilst herbivores and carnivores can be excluded or enclosed by fences and cages. Food webs are relatively simple and the limiting resources are often readily apparent. This great potentiality of rocky shores for the experimental investigation of community structure has been heavily exploited during the past two decades, with the result that more is known about the species interactions of rocky-shore communities than of almost any other ecosystem. The experimental manipulation of intertidal organisms dates back at least to the 1930s and 1940s when people like Hatton in France and Jones in the Isle of Man manipulated populations of barnacles, limpets and fucoids. Theoretical ideas on community ecology at that time, however, had not reached the critical stage necessary for these elegant experiments to blossom, and so they were not developed to their full potential. This had to wait for almost another two decades when Paine and Connell took up the reins and developed schools of research that advanced not only our understanding of rocky-shore ecology, but also of the fundamentals of community ecology itself. Paine's (1966) classical study showed how the starfish, *Pisaster ochraceus*, by thinning out populations of one mussel, *Mytilus californianus*, makes space available for colonization by inferior competitors, such as various barnacles, limpets and algae, thereby maintaining a high species diversity within the community. Others, including Dayton, Menge and Lubchenco, have extended Paine's approach and applied it to a wide range of intertidal communities and even to the sublittoral benthos under the ice sheets of Antarctica. All these studies tell fascinating stories, but just a few of those concerning rocky-shore habitats on either side of the North American continent have been chosen to illustrate general principles that have emerged.

West-coast community structure

Dayton (1971) built upon Paine's original work by examining in detail the effects of physical environmental factors and biological interactions on the upper intertidal mussel–barnacle community of the Washington coastline. The barnacle zone is colonized by three species, *Balanus cariosus*, *B. glandula* and *Chthamalus dalli*, while *Mytilus californianus* occurs in patches among the barnacles on all but the very sheltered shores and the algae *Fucus distichus* and *Gigartina papillata* occur on all but the very exposed shores. Foraging throughout the zone are four species of grazing limpet and three species of predatory dogwhelk. The large starfish *Pisaster ochraceus* moves onshore to feed during high tide.

Two important physical factors affecting community structure are wave action and battering by drifting logs (lost by the local logging industry or derived naturally from the heavily forested coastline). Drifting logs, thrown against the rocks by waves, knock off patches of sedentary organisms, making the surface available for recolonization. This effect was quantified by measuring the 'survivorship' of nails embedded into the rock with a rivet gun. Measured in this way, the probability of a patch of rock being struck by a log was well correlated with the percentage cover of mature barnacles on the shore (Fig. 5.7a). Producing a similar effect to log-battering is predation by the starfish *Pisaster ochraceus*. *Pisaster* feeds preferentially on mussels and barnacles, carrying them back down the shore if they are too big to consume during high tide. The starfish everts its stomach over the prey, and in this manner 20–60 barnacles may be removed simultaneously. Among large patches of *Mytilus californianus*, the effects of log-battering and starfish predation are complemented by wave action. After log impact, or a feeding starfish, has removed some mussels, those exposed around the edge of the clearing are twisted and torn off the rock by wave action, so that even a small clearing can be considerably enlarged.

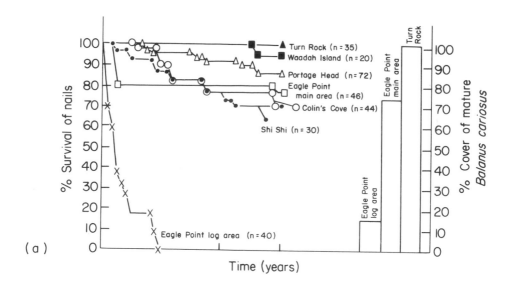

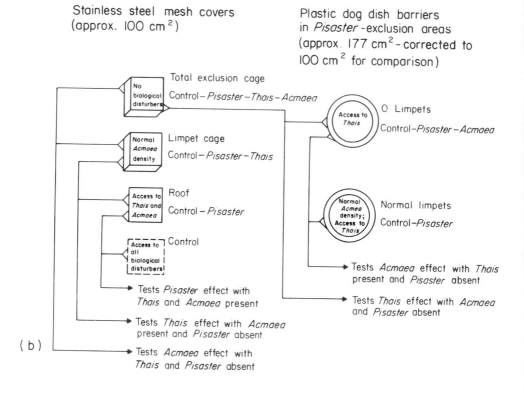

Fig. 5.7 (a) 'Survivorship' of nails embedded in rock compared with the abundance of barnacles at the same localities. (b) Experimental design used to test the effects of competition and biological disturbance (predation and grazing) in the barnacle–mussel association of a west North American shore. (After Dayton 1971.)

Two potentially important biological interactions were experimentally investigated: interspecific competition among the sedentary organisms and disturbance by the grazers and predators. An ingenious experimental design using cages and fences was made to evaluate the importances of these factors working separately and in combination (Fig. 5.7b). As a result of the experiments, the sedentary organisms could be arranged in a competitive hierarchy. *Mytilus californianus* requires intricate surfaces, provided for example by algae, barnacles, or mussel byssal threads, upon which to settle, but once established it can outcompete all the other sedentary organisms and so is the potential competitive dominant on these shores. Next in the hierarchy are barnacles, which outcompete the algae. Among the barnacles, *Balanus cariosus* is dominant over *B. glandula*, which in turn is dominant over *Chthamalus dalli*. By grazing spores and cyprids or by bulldozing newly settled barnacles off the rock, limpets reduced the recruitment of algae and barnacles in all natural situations. In the artificial situation where dogwhelks were excluded, limpet disturbance increased the survival of *Chthamalus* by reducing the survival of the competitively superior *Balanus* spp. Dogwhelks depressed the population of all barnacles in natural situations, but when limpets were artificially excluded, dogwhelks increased *Chthamalus* survival by preferentially eating the *Balanus* spp. The effect of limpets and dogwhelks acting in concert was therefore different from either acting alone, illustrating the subtlety that can occur among biological interactions and the need for experiments capable of revealing it. The competitive dominants, *Mytilus californianus* and *Balanus cariosus*, grow too big for dogwhelks to eat if they manage to escape predation for long enough. Therefore, having survived limpet grazing as newly settled spat or dogwhelk predation as young individuals, barnacles and eventually mussels would monopolize all the space in the upper intertidal zone. This is prevented by the repeated clearance of patches by log-battering and starfish predation, and the enlargement of clearings by wave action. The community structure therefore, is determined by physical and biological disturbances, which alleviate competition by creating an abundance of the potentially limiting resource—space on the rock surface (Fig. 5.8).

East-coast community structure

The relatively species-rich, mosaic structure of the mussel-barnacle zone on the west coast of North America contrasts

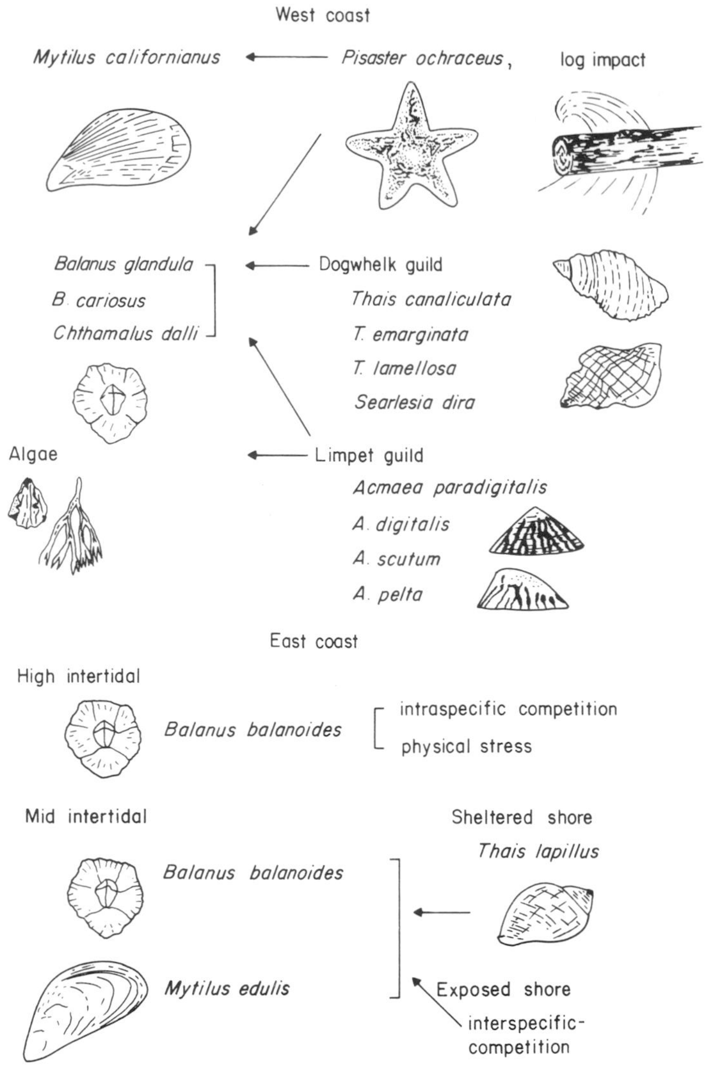

West coast

Mytilus californianus ← —— Pisaster ochraceus,   log impact

Balanus glandula ⌐ ← —— Dogwhelk guild
B. cariosus   Thais canaliculata
Chthamalus dalli ⌐   T. emarginata
                          T. lamellosa
                          Searlesia dira

Algae   ← —— Limpet guild
                  Acmaea paradigitalis
                  A. digitalis
                  A. scutum
                  A. pelta

East coast

High intertidal

Balanus balanoides ⌈ intraspecific competition
                   ⌊ physical stress

Mid intertidal   Sheltered shore
                                  Thais lapillus

Balanus balanoides

Mytilus edulis   Exposed shore
                                   interspecific-
                                   competition

Fig. 5.8 Determinants of community structure on the Washington and New England coasts of North America. (After Hughes 1980a.)

sharply with the relatively species-poor, mid- to high-intertidal zone of the east coast. The biogeographical difference seems partly due to the harsher environment (more extreme temperature fluctuations) of the east coast, which reduces the variety of species at all trophic levels (see Chapter 10). Using a similar experimental technique to that of Dayton, Menge (1976) identified the major determinants of community structure in the mid- to high-intertidal zone of exposed and sheltered rocky shores in New England (Fig. 5.8). In spite of the more austere environment, community structure on the east coast is largely determined by biological interactions, as it is on the west coast.

Grazers and community structure

Physical disturbances and predation enhance species diversity in the east- and west-coast intertidal zones by repeatedly removing significant quantities of the dominant competitors for space— mussels and barnacles. Disturbances, however, do not necessarily increase species diversity (p. 136). If too severe, as with the scraping of intertidal rock surfaces by ice in polar regions (p. 116), disturbance adversely affects all species, keeping diversity low. If disturbance is too mild to significantly affect the competitive dominants, or if it affects the subdominants disproportionately, it will not prevent the dominants from ousting poorer competitors, but may even accelerate the process, as shown by Lubchenco's (1978) experimental investigation of algal community structure in the mid–low intertidal zone of moderately exposed New England shores. The potential competitive dominants are *Mytilus edulis*, followed by *Balanus balanoides*, but these are heavily preyed upon by dogwhelks and starfish, allowing algae to predominate. Among the algae, the competitive dominants are *Chondrus crispus* (Irish moss) in the low-shore and *Fucus vesiculosus* plus *F. distichus* in the midshore region. *Chondrus* competitively excludes the fucoids from the low shore, but if kept clear of *Chondrus* the fucoids grow even better here than in their normal zone (cf. *Pelvetia*, p. 122). *Chondrus* cannot withstand desiccation at midshore levels, where the fucoids therefore have a competitive refuge. The main herbivores are winkles, primarily *Littorina littorea*, which readily consume algal spores and germlings but avoid *Chondrus*, and fucoids, that have grown beyond the germling stage. Spores that germinate in protective crevices or on smooth surfaces during the winter when winkles have migrated to the sublittoral zone, are therefore able to escape and grow into a size refuge from grazing. Occasionally *Chondrus* is removed by storms, ice-scouring, or people harvesting the plants, and the vacated rock surface is quickly colonized by ephemeral algae. If protected from grazers, the ephemerals delay recolonization by *Chondrus*, but in the natural situation *L. littorea* prefers the ephemeral algae and by preferentially removing them, accelerates regrowth of the *Chondrus* bed.

The effect of *L. littorea* grazing is quite different in rock pools high on the shore. Here *Enteromorpha* is competitively dominant over other potential colonists such as *Chondrus* and several other perennial and ephemeral algae. Lubchenco noticed that pools containing few *Littorina* supported an almost pure stand of *Enteromorpha*, whereas others containing many *Littorina* were dominated by *Chondrus* which the winkles do not eat. When *Littorina* was present at moderate densities in other pools, many perennial and ephemeral algae coexisted because the winkles preferred the competitively superior species, and reduced them to low densities without eradicating them altogether.

Species diversity therefore depends on two things. The first is whether the grazer prefers the competitively superior or inferior species. If competitively inferior species are preferred, grazing always enhances competitive exclusion, leading to spatial monopoly by the competitively superior species, such as *Chondrus crispus*, in the low intertidal zone. The second is the intensity of grazing. If competitively superior species are preferred, grazing of moderate intensity increases algal species diversity by keeping the competitively superior species at low densities. Too little grazing is insufficient to prevent the competitive dominant from ousting other species, while very heavy grazing prevents all algal germlings from growing to maturity (Fig. 5.9). The experiments and observations described above suggest that under- and over-intensive predation (grazing may be regarded as a form of predation, see p. 230) will decrease prey species diversity. Only non-selective predation (e.g. log impact) or differential predation on competitively dominant prey (e.g. starfish and dogwhelks feeding on mussels and barnacles) can promote prey species diversity. Selective predation on competitively inferior prey will decrease prey species diversity by accelerating competitive dominance.

Fig. 5.9 (a) The effect of predation intensity on species richness. (b) The relative importances of predation (solid line) and competition (dashed line) in determining community structure along a gradient of environmental harshness. (After Hughes 1980a.)

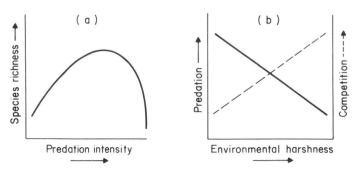

Effects of renewal of secondary space and of non-hierarchical
competition

There are two reasons why disturbances, whether they be physical or biological, are so important in maintaining species diversity on rocky shores. First, the potentially limiting resource,
space on the rock surface, is non-renewable (see section 5.1.4).
Second, the sedentary organisms form an unchanging, linear,
competitive hierarchy in which those higher up can always outcompete those below them. Left undisturbed, such a system of
competitors would always lead to monopoly by the competitive
dominant. Coexistence of competitors would be possible,
however, if either or both of the two conditions were changed.
This appears to be the case with the community of epiphytic animals living on low-intertidal fucoids.

Fucus serratus grows prolifically on moderately sheltered
British shores and supports a rich epifaunal community of
sponges, hydroids, bryozoans, serpulid polychaetes and tunicates.
Seed and O'Connor (1981) have examined the population dynamics and competitive interactions within a subset of the epifaunal
community. A crucial feature of the system is the summer
growth of the plant, which replenishes the substratum available
for colonization by up to 75% per annum (violating the first condition above). *Flustrellidra hispida* and *Alcyonidium hirsutum* are
two of the bryozoan competitors, among which *Flustrellidra* is
usually dominant. *Flustrellidra* larvae are released in spring and
summer, settling selectively on the distal growing-regions of the
Fucus serratus fronds, so producing a broad band of developing
colonies. *Fucus* continues to grow more slowly during the winter,
providing new substratum that is selectively colonized by *Alcyonidium hirsutum* larvae released in midwinter. Continued annual
cycles of *Fucus* growth and bryozoan settlement result in alternating broad and narrow bands of *Flustrellidra* and *Alcyonidium*
respectively across the distal segments of the fronds, coexistence
being made possible by the annual replenishment of unused
frond surface. Two other members of the bryozoan subset, *Electra pilosa* and *Membranipora membranacea*, are competitively inferior to *Flustrellidra* and *Alcyonidium* and survive in refuges
unused by the two latter. *Electra* and *Membranipora* settle gradually throughout the year on any part of the *Fucus* frond, making
opportunistic use of unoccupied patches.

A second important feature of the *Fucus* epifaunal community
is its competitive structure, which does not form an unchanging,
linear hierarchy, but involves a certain amount of unpredictability

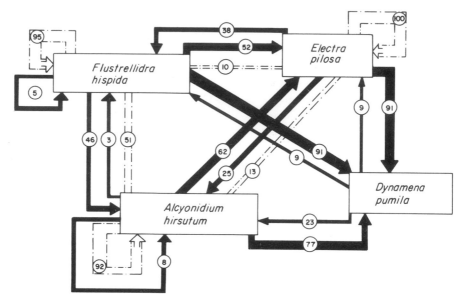

Fig. 5.10 Flow diagram of competitive interactions among four bryozoan species that colonize the fronds of *Fucus serratus*. Arrows point towards competitive subordinates. Encircled numbers are the percentage probabilities of the respective interactions taking place. (After Seed & O'Connor 1981.)

because competitive interaction can be reversed (Fig. 5.10), violating the second condition above. Although in probabilistic terms, *Flustrellidra hispida* is the potential competitive dominant, the stochastic element of the competitive interactions would prevent its monopoly of the substratum even if the latter were not replenished by plant growth each year.

Cyclical regeneration of primary space by intrinsic mechanisms

Sometimes the regeneration of bare substratum on the shore is not entirely due to external forces such as predation, storm damage or log impact, but may result from the growth activities of the competitive dominants themselves. Mussels, for example, tend to form multilayered populations as recruits attach themselves to larger residents. Founder members gradually die and sediment accumulates among the shells, as a result of which younger, actively growing mussels form 'hummocks' attached indirectly to the rock base via the layer of dead shells and sediment. As the hummocks increase in size they become mechanically less stable, eventually being washed away by waves. A mosaic of patches in various stages of maturity, from newly bared rock surface to pronounced hummocks, can therefore be maintained in a cyclical fashion.

A somewhat similar mechanism allows the small annual kelp, *Postelsia palmaeformis* (see Fig. 5.13b) to succeed in the barnacle–mussel zone of very exposed shores on the Washington coastline. Mussels and then barnacles are the competitive dominants on these shores (p. 131), but *Postelsia* patches are maintained by settlement and growth of the alga on top of mussels and barnacles. Increased drag of the growing kelp eventually causes both *Postelsia* and the animals to which it is attached to be torn away by waves. The vacated rock surface can then be directly colonized by more *Postelsia*.

5.2 Kelp forests

5.2.1 *Introduction*

Fucoid algae (Fucales) which perhaps form the most prominent feature on sheltered to moderately exposed, cool temperate, rocky shores throughout the world, are replaced at around Low Spring Tide Level (sublittoral fringe) by another group of large brown algae, the kelps (mostly Laminariales), which extend as far down into the sublittoral zone as light intensities and availability of suitable hard substrata will allow. In cool, clear waters, kelp may flourish to depths of 20–40 m and extend 5–10 km offshore on a gentle gradient.

Kelp forests are just as impressive as coral reefs (Ch. 6) in two respects: some maintain productivities of 1500–3000 g C m^{-2} yearly, comparable with the most productive aquatic or terrestrial ecosystems, while the plants themselves create an intricate three-dimensional topography (Fig. 5.11) that attracts numerous invertebrates and fish. The geographical distribution of kelp forests (Fig. 5.12) is very roughly complementary to that of coral reefs (Fig. 6.3), kelps being limited in general to latitudes between the subpolar regions and the 20°C summer isotherms and corals to latitudes within the 20°C winter isotherms. Extension of kelps into subtropical latitudes is facilitated by cold currents, usually associated with upwellings (p. 38) such as the California Current, Peru Current and Benguela Current that flow along the coast towards the equator.

Kelps alternate between an asexual sporophyte and a sexual gametophyte during their life cycle (Fig. 5.13a). It is only the sporophytes that grow into the large plants comprising the kelp

Fig. 5.11 Kelp impart an important three-dimensional structure to the habitat. Vista of Californian *Macrocystis* forest. (After North 1971.)

forest. The gametophytes are tiny filamentous plants, the ecology of which is very poorly known. The sporophyte basically consists of a holdfast, stipe and blade or lamina (Fig. 5.13b). Sometimes the whole sporophyte is ephemeral or annual, but in other cases the lamina or lamina plus stipe is deciduous and the holdfast perennial. Most of the prodigious productivity is due to growth of the blade, which occurs where the blade joins the stipe. As the blade grows it is worn away at the tip, behaving like a conveyor belt of algal tissue. Many variations exist on the basic morphological theme, ranging from simple stipes, branching stipes, simple blades, subdivided blades to the presence or absence of

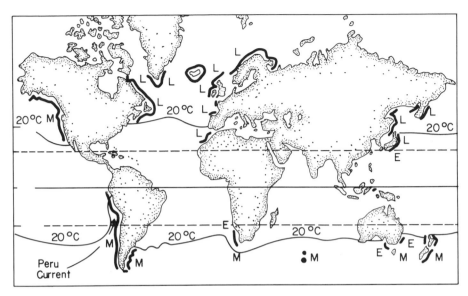

Fig. 5.12 World distribution of major kelp forests (L = *Laminaria*, M = *Macrocystis*, E = *Ecklonia*). 20°C summer isotherms are shown. (After Mann 1973.)

one or more gas-filled bladders to buoy up the blades. This diversity of form reflects the range of habitats occupied by kelps. The palm kelp, *Postelsia palmaeformis*, whose crown of multiple blades on a short, rubbery stipe, fit it for life on wave-beaten, intertidal rocks was mentioned on p. 138. Shallow, sublittoral kelps, such as *Laminaria digitata* and *L. saccharina*, common on European coastlines, tend to have short, flexible stipes and strap-like blades that are kept up in the water column by flotation and turbulence, but are able to bend limply without damage when partially emersed on low spring tides. *Laminaria hyperborea*, a dominant competitor in deeper water, has longer, stiffer stipes that lift the blades into well-illuminated water. In spite of the long stipes, *L. hyperborea* is completely submerged at high tide and this is true of all kelps in the North Atlantic. Kelp forests in this region are therefore only visible from the shore on low spring tides, when the distal regions of the stipes and pendulant blades appear just above the water surface. In the South Atlantic, Pacific and Indian Oceans, joining the strap-like laminarians is a second category of kelps typified by greatly elongated stipes bearing gas-filled bladders in the distal region. Such a kelp is like a moored buoy, the long, flexible stipe tethering the crown of blades to the sea bed, but at the same time enabling them to

141

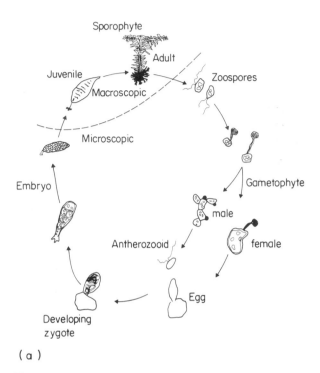

(a)

Fig. 5.13 (a) Kelp life cycle. (After North 1971.) (b) Kelp morphologies. (*Egregria laevigata* after Black R. (1974) Some biological interactions affecting intertidal populations of the kelp *Egregria laevigata*. *Mar. Biol.*, **28**, 189–98.)

remain at the surface even at high tide. Buoyed-up kelps often form dense beds parallel to the shore and visible at any state of the tide, notable examples of which are the *Macrocystis* 'giant kelp' beds off California and the *Ecklonia* beds round the Cape Province of South Africa.

Patterns and processes of community organization in kelp forests have basic features in common with those of intertidal, rocky-shore communities described in section 5.1.5. Zonation is a prominent spatial pattern and the regeneration of space on the substratum by physical and biological disturbances is an important process in both types of community.

5.2.2 Zonation

Vertical distribution

In European kelp beds, *Laminaria digitata* dominates the sublittoral fringe but is replaced by *L. saccharina* and *L. hyperborea*

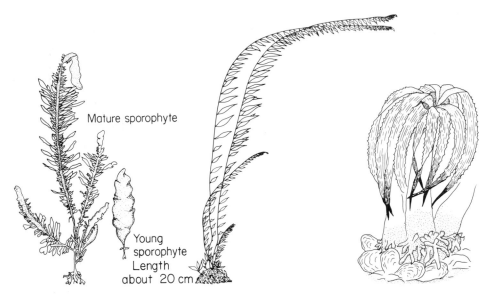

Mature sporophyte

Young
sporophyte
Length
about 20 cm

Egregia laevigata
California
Length 1−3 m

Macrocystis pyrifera
California
Length sometimes
over 50 m

Postelsia palmaeformis
West North America
Length 10−30 cm

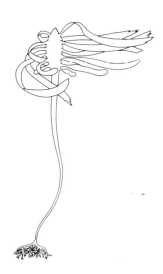

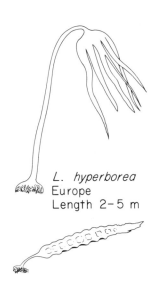

L. hyperborea
Europe
Length 2−5 m

L. saccharina
Europe
Length 1−3 m

Fig. 5.13 (b)

Ecklonia maxima
South Africa
Length 2−5 m

in deeper water in suitable localities. The upper limit of *L. digitata* is determined by an inability to survive more than the minimal tidal emersion occurring on low spring tides. Within its zone, however, *L. digitata* achieves dominance by virtue of a high growth rate persisting from late spring to late autumn (Fig. 5.14). *L. saccharina* and *L. hyperborea* are outcompeted by *L. digitata* in the sublittoral fringe because growth slows down (*L. saccharina*) or ceases altogether (*L. hyperborea*) in late summer. These different growth patterns persist in plants grown experimentally at the same depth, and are due not so much to differences in photosynthetic rate but to the way in which the photosynthates are used. *L. digitata* channels most of its photosynthates into growth right through to the autumn, so that at the beginning of the dark season, from October onwards, stored energy reserves are relatively low. In deeper water, light levels fall below the compensation point sufficiently often during winter, that stored energy reserves are necessary to sustain kelp plants until spring. Hence, *L. saccharina* and *L. hyperborea* begin to build up energy reserves in late summer, but at the expense of retarding or ceasing growth. *L. saccharina* therefore has a competitive refuge in deeper water from *L. digitata*, whose low energy reserves prevent

Fig. 5.14 Seasonal growth in three European laminarians. Holes punched in the blades move towards the tip as new tissue (black) is formed by the meristem at the base of the blade. (After Lüning 1979.)

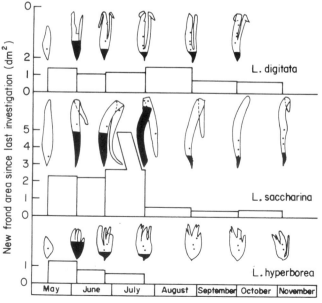

it from perennating at these depths. In still deeper water, *L. saccharina* is outcompeted by *L. hyperborea*, which achieves dominance by its long, stiff stipe and overtopping growth form. The lower limit of *L. hyperborea* is often set by grazing by the sea-urchin *Echinus esculentus*.

5.2.3   Population ecology

The population dynamics of kelps have yet to be studied comprehensively, but in spite of their size, kelps seem to have population turnover rates comparable with those of intertidal fucoids. There is a range from competitively superior, perennial species to opportunistic annuals that repeatedly invade frequently disturbed areas. Superimposed on this is the great demographic flexibility of single species, enabling many of them to flourish over a wide range of environmental conditions. Although the giant kelp *Macrocystis pyrifera* behaves as a perennial, living for several years in deeper water, its life expectancy in shallow water on exposed coasts may be less than a year due to removal by storms.

At least under the crowded conditions of kelp beds in the lower intertidal and sublittoral fringe, populations of kelp undergo the same processes of competition, self-thinning and regulation of biomass per unit area as seen in higher, intertidal, sedentary organisms (section 5.1.4). *Egregia laevigata* (Fig. 5.13b) populates the lower intertidal and shallow sublittoral zones of Californian shores. The upper distributional limit of *Egregia* on these shores is set by sun burn and sand scour, but there is extensive recruitment on exposed rock surfaces even where mature kelp never survive. The intertidal plants are annual but *Egregia* may be perennial sublittorally. A large intertidal recruitment occurs in the spring and although this is unaffected by grazers or the presence of other sedentary organisms, *Egregia* plants remaining from the previous year-class reduce recruitment by the abrading action of their stipes during the rising and falling tides. Among new recruits, survival is density dependent due to self-thinning. At high densities, holdfasts become enmeshed and some recruits are attached to the holdfasts of others rather than to the rock surface. The resulting mechanical instability causes whole clumps of recruits to detach from the substratum. By the time the fronds of the recruits begin to grow rapidly, mortality becomes density independent, but there may still be large differences in the density of recruits in different parts of the population. Biomass per unit area, however, becomes fairly even throughout the population, firstly because plants grow faster at lower densities and secondly because they develop a more branched growth form at lower densities. As a result, plant size is inversely proportional to population density.

5.2.4. *Community structure*

Grazing by sea-urchins

Often, the local dominance of a particular kelp species can be related to the degree of environmental disturbance to which it is adapted, and the overall species diversity of a kelp forest is maintained by disturbances on various scales of intensity and frequency. One kind of disturbance however—grazing by sea-urchins—is of particular importance in kelp communities from diverse parts of the world. Sea-urchins are typical members of the grazer guild in kelp forests, different species replacing each other according to geographical locality. Sometimes urchins prefer the competitively dominant kelp species and promote algal species diversity when grazing is of moderate intensity (cf. Fig 5.9). However, whatever their dietary preferences, sea-urchins in high population densities are capable of eliminating kelps and, indeed, all frondose algae, leaving a sparse-looking substratum colonized only by encrusting coralline algae, diatoms, and a close turf of filamentous green algae. For various reasons, urchin populations sometimes build up to epidemic proportions and the devastation they cause to kelp forests is reminiscent of the effect of *Acanthaster* outbreaks on coral reefs (p. 176). Prior to 1968, lush forests of *Laminaria digitata* and *L. longicruris* (= *saccharina*) dominated the rocky sea bed in a 140 km^2 bay on the outer Nova Scotian coast. The kelps extended from the infralittoral fringe to a depth of 20 m and were exceedingly productive (see section 5.2.5). The sea-urchin population of about 37 urchins m^{-2} appeared to be in equilibrium with the kelp, feeding on fragments breaking off the plants. Wave-induced swaying movement of the kelp prevented the urchins from climbing up the plants and causing damage. However, in certain places where urchins were particularly numerous, some were able to climb kelp plants, weighing them down into the reach of the other urchins. In this manner, patches several decametres in diameter were cleared within the kelp forest. Six such holes were discovered in 1968, but during the next six years they grew larger and coalesced until 70% of the kelp along a 15 km length of shore was destroyed. Complete devastation was caused as 'wavefronts' of urchins advanced on the kelp forest at rates of up to 1.7 m per month. Eventually, the entire kelp forest as such was wiped out, leaving isolated patches of kelp at the sublittoral fringe above the normal range of the urchins. Since the devastation, there has been no evidence of any recovery of the kelp forest or of a decline in the urchin population, and it remains unknown whether the devastation was a unique, but permanent, event or whether it was part of a natural cycle with a very long periodicity. A cause of the urchin outbreak has not been identified conclusively, but there is a very suggestive correlation between the urchin outbreak and decimation of the lobster population by overfishing. Lobsters feed readily on urchins and at normal densities, lobsters would be able to keep a pre-epidemic urchin population under check. Overfishing may have reduced the lobster predation pressure so much that the urchin population 'escaped' and grew into an unchecked epidemic. Restoration of the original predator–prey population densities is hindered by two phenomena. First, elimination of the kelp forest removed an important nursery ground for the lobsters and second, it would require unnaturally high densities of lobsters to bring an epidemic of urchins back to pre-existing densities.

Wide temporal fluctuations of urchin population density, however, may be typical of these animals just as they are of many benthic invertebrates with planktonic larvae (Chapter 9). Vagaries of the weather and hydrography affecting reproduction, larval survival and dispersal can cause bumper recruitment years to alternate unpredictably with long periods of almost zero settlement. In the Strait of Georgia off British Columbia, *Strongylocentrotus droebachiensis* undergoes natural periodic outbreaks that affect the kelp community, but not so drastically as did the epidemic in Nova Scotia. An increase in urchin density was triggered by a favourable plankton bloom in 1969 following an unusually long period of cool water temperatures, which extend the period of adult fertility and enhance larval survival of this cold-water species. No recruitment occurred in later years, but the urchin cohort advanced in 'wavefronts' along the coast, removing all foliose macrophytes in their path. The macroalgal community recovered its former status within four to six years. The previous urchin epidemic occurred about 20 years earlier, when sea temperatures had also been exceptionally low.

On more exposed parts of the Californian coast another urchin, *Strongylocentrotus fransiscanus*, is able to survive in shallow water cooled and aerated by wave action. The grazing pressure prevents most kelps from establishing themselves inshore, except for the annual kelp *Nereocystis leutkeana*, which is less preferred by urchins. Several hundred years ago, the inshore distribution of urchins would have been limited by the predatory sea otter, *Enhydra lutris*. After local extermination of the sea otter, urchins invaded the shallows and eradicated most of the kelps. In order to boost crops of the commercially exploited giant kelp, calcium oxide is now used in some areas to poison urchins, in place of sea otter predation.

Role of the sea otter

Originally, the sea otter was a 'key-stone' species of nearshore communities from the northern Japanese archipelago, through the Aleutian Islands, down the Pacific coast of North America to Baja California, but was almost extinguished from the whole of this geographical range by fur traders. Sea otters have now been re-established in many areas, where their effect on the shallow benthic communities has become very evident.

At Amchitka Island in the Aleutians for example, sea otters now occur at a density of 20–30 km^{-2} and consume about 35 000 kg km^{-2} yearly of prey, mainly urchins, molluscs, crabs and fish. Sea otters do not forage effectively below 18–20 m and at these depths *Strongylocentrotus droebachiensis* becomes very abundant and eradicates nearly all frondose algae. Here, the substratum is covered by encrusting coralline algae and close-growing green algae. Around Attu Island, where sea otters are absent, a similar sparse algal flora extends to the sublittoral fringe, but at Amchitka Island sea otters remove nearly all urchins from the shallow water and also reduce the densities of sedentary animals such as mussels and barnacles. Freed from heavy grazing pressure and from intense competition with barnacles and mussels, a very dense kelp flora is able to flourish within the foraging depth-range of the sea otter. So productive is this lush algal vegetation (1275–2840 g C m^{-2} yearly) that it once supported Stellar's sea-cow, *Hydrodamalis gigas*, a giant (*c.* 10 t) sirenian, long ago exterminated by hunters. It seems likely that the sea otter was indirectly responsible, by checking populations of invertebrate grazers and competitors, for the high productivity of large algae necessary to maintain the sea-cow populations (Fig. 5.15).

Competition in an Aleutian kelp community

The community dynamics of the species-rich Amchitka kelp forests have been elucidated by Dayton (1975), using some of the

Fig. 5.15 Stellar's sea-cow, *Hydrodamalis gigas*, used to browse on North Pacific kelps. The rich supply of kelp necessary to support the sea-cow may have been maintained by the sea otter, *Enhydra lutris*, which feeds on sea-urchins and molluscs that will eradicate kelps if left unchecked. (After Scheffer 1973.)

experimental techniques previously applied to intertidal communities (section 5.1.5). The algal community has four separate canopies (Fig. 5.16). *Alaria fistulosa* has long, floating blades that form a canopy on the surface and is most dense at depths less than about 5 m. The second canopy level is formed by *Laminaria groenlandica*, *L. dentigera*, *L. yezoensis* and *L. longipes* and occurs from the sublittoral fringe to a depth of about 20 m. The third canopy is formed by *Agarum cribrosum*, whose broad fronds supported by a short stipe lie prostrate on the sea bed. The fourth canopy is a turf of numerous species of red algae together with occasional green algae. In shallow (< 10 m) water, the canopies

149

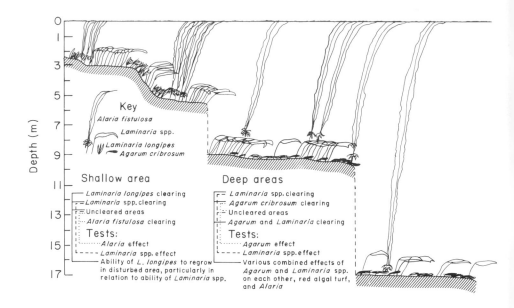

Fig. 5.16 Canopies at three different depths in an Aleutian kelp forest. The experimental design used to investigate the effects of competition among algae is also depicted. (After Dayton 1975.)

tend to occupy non-overlapping patches, suggesting the effect of interspecific competition. The experimental design used to test this supposition is summarized in Fig. 5.16. From the experiments it became clear that in spite of forming a surface (primary) canopy, *Alaria fistulosa* is not a competitive dominant, but an opportunistic species, colonizing shallow areas cleared of the dominant *Laminaria* by storms. Among the laminarians however, *L. longipes* has a rhizoidal growth pattern enabling it to regenerate very quickly after its canopy has been torn away and so this species tends to retain its local monopoly of the substratum. In shallow water, *Laminaria* spp. overshadow and outcompete *Agarum cribrosum*, but in deeper water where sea-urchins are moderately abundant, *Agarum* becomes more successful because it is distasteful to the urchins and is able to colonize areas where urchins have removed *Laminaria*. The *Agarum* (tertiary) canopy prevents the recruitment of *Alaria*, which in deeper water can only survive by attaching high on *Laminaria* stipes. The secondary and tertiary canopies suppress the growth of the red algal turf, which opportunistically invades newly cleared patches along with *Alaria*.

5.2.5 *Productivity*

Kelps often live in an environment ideal for plant growth. Photosynthesis is neither limited by desiccation nor by excessive insolation. Wave action keeps the blades in constant motion, providing maximum exposure to sunlight and enhancing the uptake of nutrients. The overall supply of nutrients is kept replenished by wind-induced mixing and upwelling of deeper water. Kelp forests in Nova Scotia maintain a productivity of about 1750 g C m^{-2} yearly compared with a productivity of 640–840 g C m^{-2} in the *Fucus*- and *Ascophyllum*-dominated intertidal zone. What happens to this prolific production of organic matter? Under certain circumstances a good proportion of it may be consumed by herbivores. Until exterminated in the eighteenth century, the giant Stellar's sea-cow browsed the lush kelp forests of the North Pacific coastlines (p. 149) and in California the giant kelp *Macrocystis pyrifera* provides a commercial harvest of 10 000–20 000 t (dry weight) yearly. In most circumstances, however, it appears that the majority of kelp production enters the community food web via the detritus food chain, just as does primary production in mature terrestrial ecosystems. Some of the detritus is derived as fragmented particles, but a large proportion is derived from the release of dissolved organic matter that is flocculated by bacteria.

South-African kelp forest

South-African kelp forests contain two dominant species, the buoyed-up *Ecklonia maxima* predominating in turbulent inshore areas, and the subsurface *Laminaria pallida* continuing into deeper water. Fig. 5.17a summarizes the community energetics of a kelp forest on the Atlantic coast of the Cape Province. Wind and waves supplement the energy input from the sun. South-easterly winds cause the upwelling of nutrients and waves keep water flowing over the kelp blades, thereby maintaining concentration gradients facilitating the supply of nutrients and removal of wastes at the kelp surfaces. Kelp production amounts to about 25 900 kJ m^{-2} yearly, about 15% of which is represented as plants uprooted by storms. Uprooted *Ecklonia* floats and is washed out to sea or cast onto the shore, where it is either collected for commercial use or is consumed on sandy beaches by teaming populations of the amphipod *Talorchestia capensis*, or on rocky shores by the isopod *Ligia dilatata*. Faeces of these crustaceans, together with any remaining kelp fragments, enter the detritus food chain either intertidally or as particles suspended by wave action and consumed sublitorally by filter-feeders.

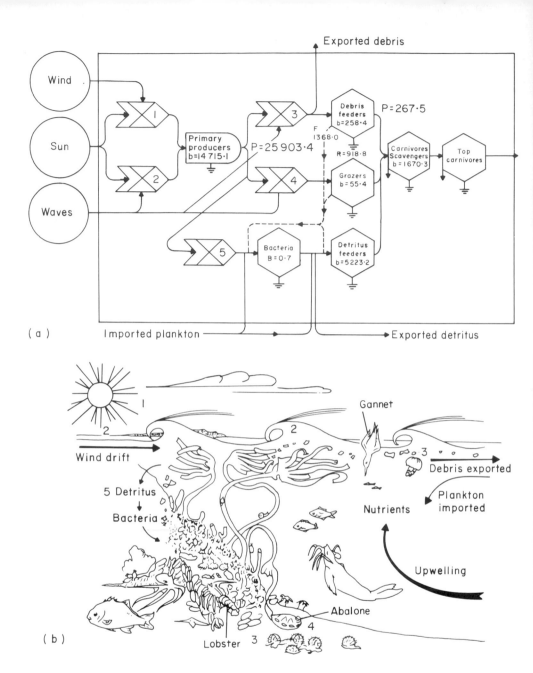

Fig. 5.17 (a) Energy circuit diagram for a South African kelp community. b = standing crop, B = standing crop in water column under 1 m^2 sea surface, P = production, R = respiration, F = faeces. (All units are kJ m^{-2}.) The numbered work gates are: (1) south-west winds causing upwellings; (2) waves moving water near the kelp surfaces; (3) waves uprooting kelp; (4) waves keeping the kelp in motion, preventing grazers climbing them but also bending them within reach of the abalone; and (5) waves eroding the tips of the kelp blades. (From Velimirov *et al.* 1977.) (b) The effects of sun, wind and waves in driving a kelp bed ecosystem. The numbers refer to the work-gates depicted in (a). Waves erode the kelp blade-tips, mussels and holothurians filter detritus from the water, sea-urchins feed on broken pieces of kelp and abalones trap kelp blades underfoot in order to graze them. (After Field *et al.* 1977.)

Uprooted *Laminaria* sinks to the bottom, remaining within the kelp forest where it is consumed mainly by the sea-urchin *Parechinus angulosus*. Waves continually erode the tips of kelp blades as in the Nova Scotian kelp forests, releasing small particles and dissolved organic matter that are utilized by bacteria and converted into detritus particles, which are in turn consumed by filter-feeding and deposit-feeding benthic animals.

Energy subsidies from winds even affect the bacterial 'machinery'. Bacterial biomass in kelp forest sea-water is minimal during winter, but south-easterly winds during summer generate upwellings with high densities of bacterial aggregates in the water. In addition to fragmentation, the kelps produce mucilage at a rate of about 16 700 kJ m^{-2} yearly, supporting an annual dry biomass of bacteria of some 43 g m^{-2}, which in turn supports an annual dry biomass of flagellates and ciliates of about 4.3 g m^{-2}. The bacteria and protozoa are probably consumed directly by the filter- and deposit-feeding benthic invertebrates. In addition to its important effects on primary production and on the detritus food chain, wave action also affects the grazers. By keeping kelp plants in perpetual motion, wave action prevents grazers such as urchins and the large trochid snail *Turbo cidaris* from climbing up the plant and damaging the canopy. On the other hand, some plants are bent over so much by waves that the blades sweep over the sea bed where the abalone *Haliotis midae* can trap them underfoot. The main carnivore is the rock lobster *Jasus lalandii* that feeds heavily on mussels, but also takes a variety of other invertebrates, and is in turn eaten by dogfish, seals, octopus and cormorants. These trophic relationships are summarized in Figs. 5.17a, b, from which it is clear that the detritus food chain accounts for the great majority of kelp primary production. The South African study emphasizes particularly well the importance of energy subsidies in the form of wind and waves, which enable kelps to make maximum use of the input of solar energy and sustain productivities rivalling those of coral reefs, tropical rain forests or even the most intensive agricultural ecosystem.

6 Coral Reefs

6.1 The coral organism

Coral reefs are a naturalist's paradise. Diving among them is like entering another world: clear, warm waters allow one to drift comfortably over vistas of intriguing growth forms, among which swarm a kaleidoscope of brightly coloured fish. Even sound may augment the visual impact, for on diving very close to the substratum, auditory signals of multitudinous pistol-shrimps hiding within crevices can often be heard like the crackle of a breakfast cereal. The vast quantities of limestone produced by corals, their high productivity in otherwise unproductive seas, the richness of species and subtlety of biological interrelationships engendered by them, put coral reefs at the centre of attention of numerous scientists from geologists and biochemists to biologists.

What is the structure of the organism that has such a profound impact on nature? Corals are closely related to sea anemones, and may be visualized as colonial anemones that secrete limestone foundations providing structural support and protection (Fig. 6.1). Each polyp sits in a cup-like depression, the calyx, which has radiating fins projecting from the base. In times of danger or physical stress, the polyp contracts so that the tissue is crammed between and over the fins within the calyx. At other times the polyp extends to varying degrees, depending on time of day and on species. Polyps are connected to their neighbours in the colony by a thin sheet of tissue, the coenosarc, covering the limestone between the calyces. Polyps and coenosarc therefore constitute a thin, essentially two dimensional, layer of living tissue over the block of limestone they have secreted. The mouth is centred within a crown of tentacles, which may be well developed or barely discernible according to species. The stomach is a simple sac with numerous longitudinal folds, or mesenteries, directed towards the centre. The radial plates of the calyx project upwards between the mesenteries. The free edge of each mesentery contains enzyme-secreting cells and is greatly convoluted towards the base of the mesentary, forming mesenterial filaments that may be extruded from the mouth or through temporary holes in the body wall for extracoelenteric digestion or defence (see Fig. 6.11). Multitudes of symbiotic, unicellular algae, the zooxanthellae, pervade the coral tissue, imparting to it a dull brown, green or blue colouration. Zooxanthellae in all corals apparently belong to a single species of dinoflagellate, *Symbiodinium microadriaticum*, different strains of which are specific to particular host-species. Without the zooxanthellae, corals become pallid, and whereas corals provide much of the topographical richness of

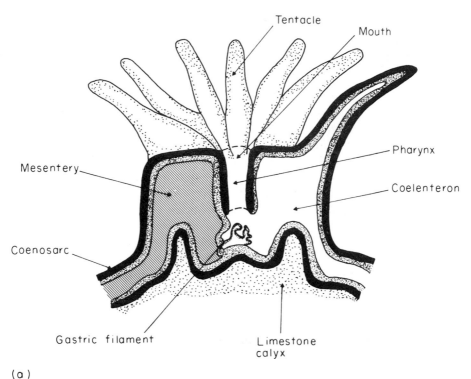

(a)

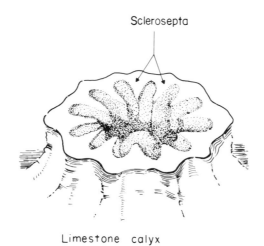

(b)

Fig. 6.1 (a) Coral polyp, showing arrangement of tentacular crown, pharynx, coelenteron ('stomach'), mesenteries, mesenterial filaments and coenosarc. (b) Calyx secreted by the polyp and upon which the polyp rests, showing arrangement of vertical septa that project into furrows in the wall between the mesenteries of the polyp.

155

the reef, other organisms, notably gorgonians, sponges and fish, provide most of the colour. Nevertheless, the sombre hues of the zooxanthellae signify a crucial biological function, for it is the mutual exchange of algal photosynthates and coelenterate metabolites that provides the key to the prodigious biological productivity and limestone-secreting capacity of reef corals.

Although all corals secrete calcium carbonate, not all of them produce sufficient quantities to form reefs. Warm seas and high light intensities throughout the year are necessary for the algal–coelenterate symbiotic machinery to produce copious quantities of limestone. Reef-building (hermatypic) corals are therefore restricted to within about 70 m of the surface in clear seas where the temperature remains above $20°C$ throughout the year. In colder regions, murkier waters, or at depths below 70 m, corals may still exist on suitable hard substrata, but although the symbiotic relationship with zooxanthellae often persists wherever there is sufficient light for photosynthesis, the capacity of the corals to secrete limestone is greatly reduced, so that they form relatively small, flimsy, colonial skeletons or even exist as solitary polyps like the Devonshire cup coral (Fig. 6.2). Solitary corals, e.g. *Fungia*, also exist on coral reefs but they tend to be large, massive species. How zooxanthellae accelerate calcium carbonate deposition is not fully understood. An early idea hypothesized the following reaction:

$$Ca^{2+} + 2HCO_3^- \rightleftharpoons Ca(HCO_3)_2 \rightleftharpoons CaCO_3 + H_2CO_3 \qquad (6.1)$$

In the light, photosynthesis is supposed to accelerate calcium carbonate deposition by using up carbon dioxide (as carbonic acid), thus shifting the equilibrium to the right. This hypothesis, however, cannot explain why the apical polyps of acroporid corals calcify much faster than lateral ones even though apical polyps contain fewer zooxanthellae. Other untested hypotheses are that photosynthates might be used in the synthesis of the skeletal organic matrix (translocation of photosynthates towards the apex of a branch might explain the faster calcification by apical polyps of acroporid corals) and that the uptake of phosphate waste metabolites by zooxanthellae removes these potential

Fig. 6.2 Coral colonies of different morphologies (not to scale): (a) Devonshire cup-coral, a temperate, non-hermatypic species; (b) massive coral with partially separate polyp heads; (c) massive 'brain' coral; (d) massive lobed coral; (e) massive plate-like coral; (f) foliose bracket-like coral; (g) small branched coral; (h) branched 'stag horn' coral; and (i) branched 'elk horn' coral.

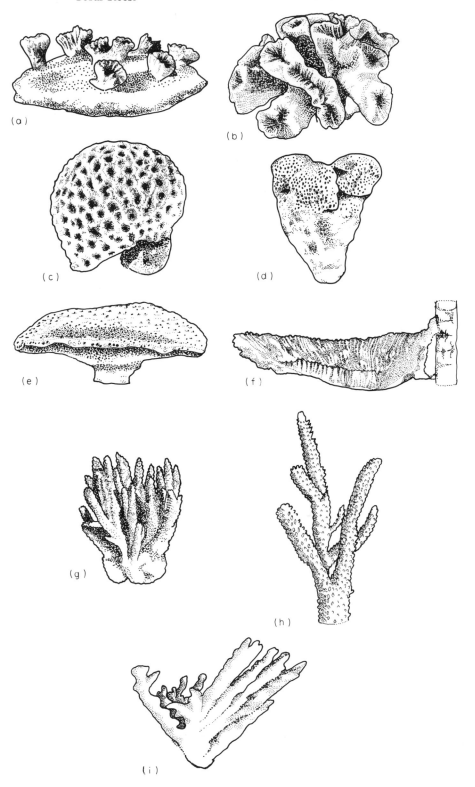

crystal poisons from the calcifying microenvironment, perhaps explaining why in darkness corals with zooxanthellae calcify faster than those without.

6.2 The structure of reefs

Charles Darwin in 1842, produced the first distributional map of coral reefs throughout the world. On it, he distinguished the three main geomorphological categories of reef still recognized today: fringing reefs, barrier reefs and atolls. Much more has been learned about the geography of coral reefs since Darwin's time, but he was correct in the essentials. Fig. 6.3 summarizes the worldwide distribution of living reefs, which can be understood in terms of suitable temperature regimes, water clarity,

Fig. 6.3 World distribution of coral reefs. (After Schumacher 1976.)

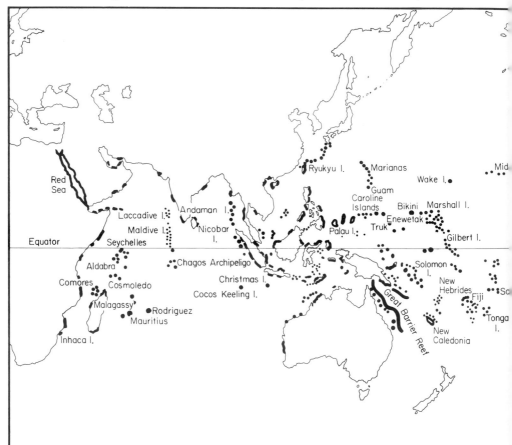

salinity and geological foundations. In the Atlantic for example, coral reefs are confined to the western parts of the ocean. Off the west coast of Africa, conditions are made unsuitable for reef formation by seasonal influxes of the cool Guinea Current, local upwellings of cold water and large amounts of silt deposited by rivers. In the western Atlantic, silt deposited by the Orinoco and Amazon rivers curtails reef formation south of the Caribbean, while north of southern Florida, low winter temperature is the limiting factor. Between these latitudes, reef formation is limited in the Gulf of Mexico by turbidity and lack of rocky foundations. Certain corals can grow in turbid waters and may form discontinuous patches on sedimentary substrata, but fully developed reefs are not formed in these situations. Optimal conditions for reef corals are to be found in the Red Sea, where the surrounding

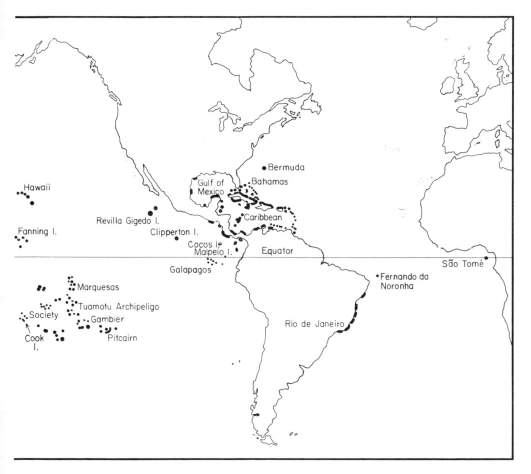

arid terrestrial climate minimizes silting and dilution from runoff, and the shape of the basin prevents upwelling or the intrusion of cold oceanic currents. Elsewhere in the vast Indo-Pacific region, reef corals tend to be most prolific around offshore islands, oceanic islands or along mainland coasts such as East Africa and eastern Australia, where terrestrial runoff is low.

Fringing reefs are formed close inshore on rocky coastlines by the growth of corals and associated hydrozoans (stinging corals), alcyonarians (soft corals) and calcareous algae. The non-coral organisms are always important, and in some situations more important, than the corals themselves. Fragments of limestone derived from these reef-building organisms (bioherms) are welded together by encrusting calcareous algae and by the deposition of interstitial calcium carbonate cement, brought about by geochemical reactions and perhaps by bacterial action. Biohermatypic accretion is entirely seaward on expanding reefs if water depth remains constant or decreases over geological time, vertical growth being curtailed at about Low Spring Tide Level by the inability of corals to withstand prolonged emergence. The

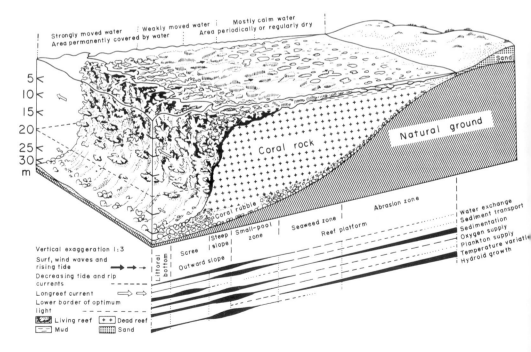

Fig. 6.4 Fringing reef of the Red Sea. Accretion of limestone proceeds seawards, leaving an ever-widening reef flat between the reef edge and the shore. (After Mergner 1971.)

zone of living corals is usually separated from the shore by a shallow reef-flat, where reduced water circulation, periods of tidal emersion and the accumulation of sediments makes conditions unsuitable for prolific coral growth (Fig. 6.4).

Vertical accretion takes place over geological time if water depth increases due to subsidence of the sea bed or to a eustatic rise in sea-level (i.e. relative to the earth's centre). During the last sequence of ice ages there were alternating periods of high and low sea-levels. Vertical reef growth would have been accordingly intermittent, occurring only at the highest interglacial sea-levels. Low glacial sea-levels would have exposed the reefs to the air, subjecting them to erosion. We now live in an interglacial period with a rising sea-level, so that a general vertical component to biohermatypic accretion, in the order of 3–15 mm yearly, has been maintained over the past ten thousand years. Rapid coral growth is restricted to within about 20 m of the sea surface, being most rapid on the upper slope just below the reef crest.

Barrier reefs are separated from the land by a lagoon, usually formed by coastal subsidence, during which only the seaward

Fig. 6.5 Atoll formation by the subsidence of a volcanic island and the compensatory vertical growth of coral.

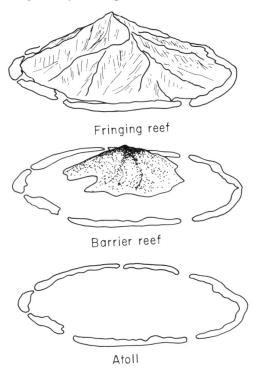

Fringing reef

Barrier reef

Atoll

reef, by continued upward growth, is able to maintain contact with the sea surface. Atolls are simply annular reefs formed round subsiding volcanic islands (Fig. 6.5), an explanation first given by Darwin who was impressed by the force of tectonic earth movement revealed in the uplifted rock strata of the South American Andes.

There are gradations between the three types of reef and variations on the basic theme. The Great Barrier Reef off the Queensland coast of Australia for example, is not a simple barrier reef as defined above. It is in fact the largest aggregation of reefs in the world, covering an area of over 80 000 square miles, and is composed of many reefs differing in shape and size.

Fringing reefs, barrier reefs and atoll reefs have the same basic biological structure and processes of accretion. A seaward transect across any reef would reveal an ordered sequence of geological and biological structure. As with any environmental gradient, whether down the shore, along an estuary, or up a mountainside, organisms are adapted to different positions along the gradient so that they follow a recognizable, albeit extensively overlapping, pattern of zonation (p. 117). On coral reefs, this zonation is most pronounced on very exposed windward reefs and least pronounced on sheltered leeward reefs. The reef flat is usually backed by a sand beach or, as in the Red Sea, East Africa, and many Indo-Pacific Islands where old reefs have been raised tectonically, by low limestone cliffs. Accumulation of sediments, shallow water and frequent emersion, prevent corals from growing extensively in this zone. Hard substrata are dominated by frondose algae and sediments by sea-grasses, both harbouring a rich fauna of decapod crustaceans, molluscs and worms. Further seaward, slightly deeper water and greater mixing by waves spilling over the reef create an environment more favourable for corals and alcyonarians, which gradually replace the frondose algae. Corals tend to be small, branched forms that are able to grow quickly in bright light and to repopulate areas of reef flat devastated by storms or by exceptionally low tides (p. 178).

The structure of the reef crest and seaward reef slope differ considerably between windward and leeward reefs (Fig. 6.6). On very exposed, windward reefs, storm waves pile up dislodged coral heads and rubble forming a boulder or rubble zone parallel to the reef, often associated with gravel tongues extending at right angles onto the reef flat. Persistent breakers prevent sediment accumulating seaward of the boulder zone, which may therefore be separated from the reef crest by a shallow moat. The reef crest is of slightly higher relief, reaching just above Low

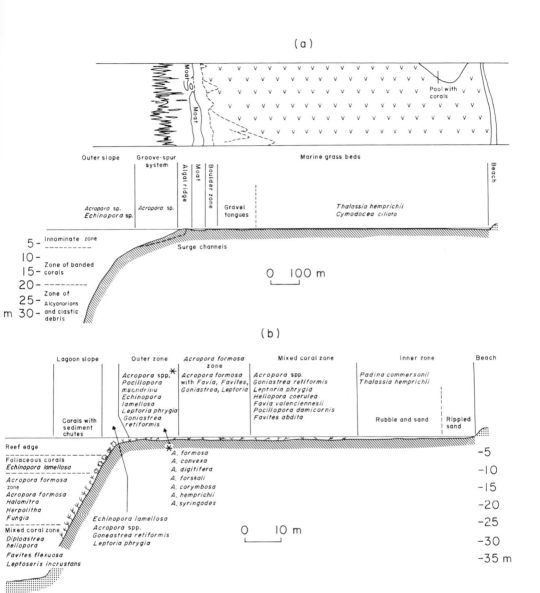

Fig. 6.6 Structural features of: (a) the windward reef; and (b) the leeward reef. (After Stoddart 1971.)

Spring Tide Level. For reasons not yet fully understood, different organisms dominate the reef crest in different geographical areas. In places exposed to the most severe wave action, the reef crest is dominated by encrusting calcareous algae forming an algal ridge. This feature is widespread throughout the Indo-Pacific but is less common in the Caribbean. Although exposed on the lowest tides, each breaker sends a sheet of water over the ridge, preventing desiccation. Accretion of the ridge is almost entirely due to the growth of encrusting calcareous algae, predominantly *Porolithon*, that form an internally honeycombed limestone with a smooth outer surface that is able to absorb and deflect the enormous energy of persistent swell. Small crabs, shrimps, cowries and other animals reside in the multitudinous subsurface cavities of the algal ridge, protected from waves and predators. Parts of the algal ridge may be formed of dense colonies of vermetid gastropods, whose upwardly growing tubular shells are cemented together by the calcareous algae. In certain places, especially near the edges of drainage channels, there may be carpeting clones of zooanthids (colonial 'sea anemones') whose brown or green tissues feel soft and slippery underfoot. Sea-urchins of the genera *Echinometra* and *Echinostrephus* often occur in dense populations, each urchin excavating a depression for itself in the calcareous substratum.

In situations of moderately severe wave action, windward reef crests tend to be dominated by one or two coral species, notably stoutly branching corals such as *Acropora palmata* in the Atlantic, *A. cuneata* in the west Pacific, *Pocillopora* spp. in the east Pacific, or by robustly branching 'stinging corals', *Millepora* spp., on reefs in various parts of the world. The close-growing, robust colonies form ramparts able to withstand the heavy seas. Both algal- and coral-dominated reef crests tend to be dissected by a 'spur-and-groove' system normal to the reef edge and extending some distance down the reef slope (Fig. 6.7a, b). The alternating spurs and grooves are several metres wide and may be up to 300 m long depending on the slope of the reef. They appear to be formed by erosion reinforced by the prolific seaward growth of corals on the grooves. The relative importances of differential erosion and coral growth remain unknown, but development of the spur-and-groove system is evidently self-reinforcing. Incoming waves cause tremendous surges back and forth along the grooves, churning up sediment and rolling boulders so that coral growth is prevented. Less turbulent water on the spurs allows corals to grow, reducing turbulence further. Sometimes corals from neighbouring spurs arch over the intervening grooves, forming canyons or even tunnels leading to blow holes on the reef crest. The effect of the spur-and-groove system is to dissipate the tremendous force of the unabating waves, estimated to be about half a million horsepower at Bikini Atoll, so stabilizing the reef structure. So severe and persistent is the water motion in the first few metres down the reef slope, that diving or sampling of any kind is virtually impossible most of the

Coral Reefs

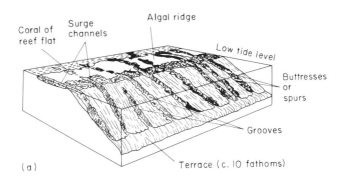

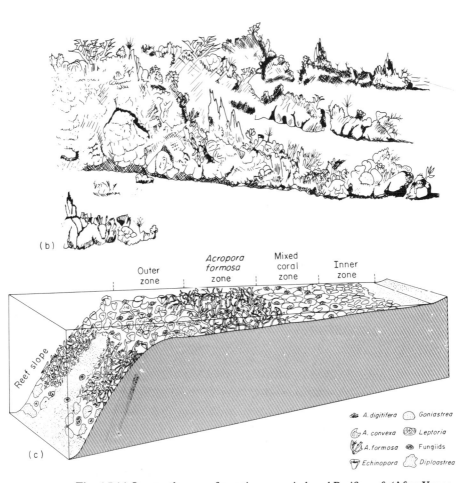

Fig. 6.7 (a) Spur-and-groove formation on a windward Pacific reef. (After Yonge 1963.) (B) Butresses (equivalent to spurs) of a Caribbean reef. (After Goreau & Goreau 1973.) (c) Coral zonation on a lagoon (sheltered) reef of a Pacific atoll. (After Spencer-Davies *et al.* 1971.)

165

time, and for this reason the uppermost part of the reef slope has been termed the innominate zone (Fig. 6.6).

At greater depths the force of water movements is less, but so also is the light intensity. The less firmly cemented reef structure and the prolificacy of sponges, molluscs and worms that excavate and weaken coral skeletons at these depths, increase the risk of dislodgement by collapse and slumping of the substratum. Coral growth-forms therefore tend to be more foliaceous or plate-like, with large areas for the interception of light from above and with hydrodynamic properties tending to prevent them from resting wrong-side-up after dislodgement or from rolling down the reef slope into lethal depths. Another, not necessarily mutually exclusive, explanation for the change from massive to foliaceous growth forms at greater depths is that diminished light intensities restrict the photosynthesis of zooxanthellae, and hence the rate of calcium carbonate deposition. Since polyps can feed on plankton, which does not decrease in abundance with depth, the growth of coral tissue outstrips skeletal growth, resulting in flattened, plate-like colonies. Contrary to this idea, however, the foliaceous coral *Agaricia agaricites* produces calcium carbonate more slowly in relatively shallow water.

Sponges, alcyonarians (seawhips and gorgorians) and non-hermatypic corals become increasingly important down the reef slope and gradually replace hermatypic corals below 30–70 m depth, depending on water clarity. Reduced water movements allow more silting and the substratum below the talus slope usually consists of fine sediments.

On leeward reefs zonation along the reef flat is essentially similar to that on windward reefs, but the boulder zone, moat, algal ridge and spur-and-groove system are absent (Fig. 6.7c). Coral associations on the outer reef flat merge gradually with those on the upper reef slope. Diverse forms are present, but there is an abundance of branching, quick-growing forms. Corals tend to be aggregated into patches or large mounds, between which are pockets of sand and rubble (Fig. 6.7c). The topographically diverse habitat supports a rich invertebrate and fish fauna, notable members of which are the black sea-urchins (*Diadema* spp.) whose needle-like spines readily pierce unwary feet.

Changes in coral morphology with increasing depths down the reef slope are similar to those on the windward reefs, and in both cases are due more to species replacements in the Indo-Pacific and more to phenotypic plasticity in the Caribbean. On the slopes of some Pacific reefs, for example, species of *Pocillopora*, *Porites* and *Pavona* gradually replace each other with increasing

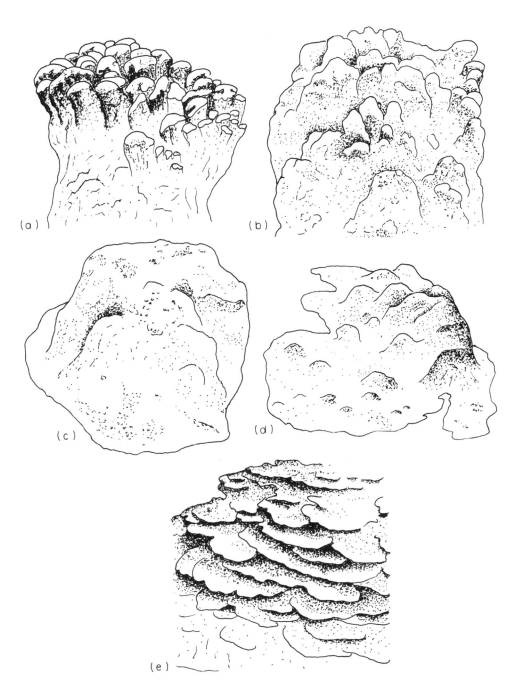

Fig. 6.8 Morphological changes in the massive Caribbean coral *Montastrea annularis* growing at increasing depths: (a) 5 m; (b) 13 m; (c) 18 m; (d) 25 m; and (e) 35 m. (After Barnes 1973.)

depth, whereas on Caribbean reefs the entire depth range is covered by fewer species, which respond to different light levels by altering their morphology. *Montastrea annularis* thus changes from a stoutly branched, to a massive and finally to a foliose morphology with increasing depth (Fig. 6.8).

Not only morphology, but also coral species diversity changes with depth (Fig. 6.9). The wave-battered reef crest and upper reef slope are colonized by a few robust species, described previously. Within a few metres down the reef slope, species diversity increases in response to the ameliorating conditions. Optimal light intensities and the reduced effect of wave action allow the maximum species diversity to develop round about 20 m depth. At greater depths, the attenuation of light causes species diversity to decline.

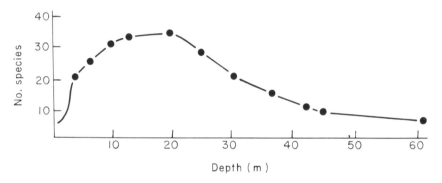

Fig. 6.9 Number of coral species present at increasing depths on an Indian Ocean coral reef. (After Sheppard 1980.)

Biohermatypic accretion is quicker on sheltered than on exposed reefs. In recent geological time, relatively dense limestone has accumulated at a rate of about 3–6 m per 1000 years on high-energy reefs, whereas on more sheltered reefs dominated by acroporid corals, less dense limestone has accumulated at a rate of about 8–15 m per 1000 years.

6.3 The life histories of corals

Coral reefs are often regarded as equivalent to tropical rain forests in terms of their species richness. The comparison is a good one, because the potential longevities of corals and the factors determining their recruitment or replacement turn out to be rather similar to those for trees in certain rain forests. The immediate neighbours of a coral colony tend to be of different species. Why do the faster-growing species not oust the slower growing ones? In certain situations they do, as evidenced by the thickets of branching corals frequently found on the upper slopes and crests of moderately exposed to sheltered reefs (Fig. 6.7c). How do these monospecific stands originate, and what limits their spread? To answer these questions, we need to understand the life histories of corals, their modes of recruitment, growth, and reactions to competition, in addition to the nature and frequency of disturbing factors such as predation, excavation by boring organisms, or dislodgement and fragmentation by storms.

Reproduction and dispersal of corals

The life history of corals is typical of sedentary animals: dispersal is primarily by tiny planktonic larvae called planulae produced in large numbers to compensate for the hazards of passive transportation in water currents (see Chapter 9). After its relatively brief, motile existence, the planula settles and metamorphoses into a polyp. Upon growing to a certain size the polyp divides, this process continuing for the rest of the animal's life and forming an ever-expanding clone. The clonal colony becomes sexually mature at some minimum size, about 10 cm diameter, corresponding to an age of about eight years in the massive coral *Favia doreyensis*, but probably occurring in younger colonies of the faster-growing, branching corals such as species of *Acropora*, *Pocillopora* and *Stylophora*. Gonads form on the mesenteries, some species being simultaneous hermaphrodites, others either sequential hermaphrodites or gonochoristic (separate sexes). Cross-fertilization is thought to be the rule, sperm liberated in the sea-water being drawn by ciliary currents into the coelenteron of another clone where the ova are fertilized on the mesenteries. Reproduction is seasonal in some places and species, but continued throughout the year in others. As in forest trees, individuals do not necessarily breed every year, and this is more likely to be the case in the slow-growing, long-lived corals than in faster-growing, shorter-lived species.

The number of eggs produced per polyp will follow the general inverse relationship between egg size and number, larger eggs being associated with a shorter larval period and reduced potential for long-distance dispersal (p. 240). Total fecundity is determined by the number of polyps and will continue to increase with colonial growth. A moderately sized colony may be expected to produce up to several thousand planulae a year.

Newly released planulae swim upward (geonegative) and towards the light (photopositive) so that they enter the surface waters and get carried along by the current. Planulae of most species studied reverse their response to gravity and light within two days, swimming to the bottom where they settle. Others remain swimming for three weeks to two months and are capable of very long-distance dispersal. Nonetheless, the majority of planulae contributing to the normal recruitment on a reef, settle within about two days of release and must, therefore, originate from corals in the same area. Even though planulae may not be carried great distances in two days, it is unlikely that they would end up near to, and compete as new colonies with, their parents. Owing to larval mortality, the settlement rate will be considerably less than the rate of planular production. After settlement, mortality rates will be less, but still quite high until the colonies have increased in size, thereby gaining some immunity to risks of predation or mechanical damage. From sequential counts of colonies about 1 cm in diameter and therefore approximately a month old, Connell (1973) estimated that about five young colonies m^{-2} recruited yearly on to a reef flat on the Great Barrier Reef. Half of these young recruits died within a year.

Growth and longevity of corals

Space in the light is a critical resource for hermatypic corals. Calcification rates on sunny days can be double those on cloudy days in shallow-water forms, and experiments cutting out 90% of the incident light have killed such corals within two to six months. Although seasonal fluctuation in day length and water temperature are only slight in the tropics, seasonal growth pattern in corals are evident in X-radiographs of the skeleton (Fig. 6.10). Alternating light and dark bands correspond to periods of faster growth when a less dense limestone is produced, and of slower growth when denser material is laid down, each pair of light and dark bands representing one year's growth. At Enewetak Atoll in the Pacific, high-density bands are secreted from about July to January, when light intensity and temperatures are higher, and low-density bands are secreted during February to June when light intensity and temperatures are lower. The high-density band forms a larger proportion of the annual growth increment in slower-growing corals, thus accounting for the denser limestone derived from such species (see p. 168).

Annual growth rings provide a record of past growth rates and so enable comparisons to be made that might otherwise take

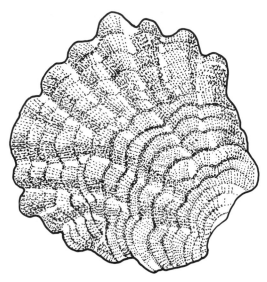

Fig. 6.10 X-radiograph of a section of the skeleton of *Porites lutea*. Light bands indicate the faster production of lower-density limestone and dark bands the converse. (After Highsmith 1979.)

years to complete. Examination of several species of Enewetak corals has revealed no sign of a decline in growth rate with ages of up to 30 years. Most animals could not increase in size indefinitely and still maintain a constant growth rate. Numerous invertebrates continue to grow throughout life, but for many reasons growth slows down towards an asymptote in old age. One potential factor limiting growth rate is the decreasing surface area to volume ratio that accompanies an increasing bodily volume. Some corals perhaps escape this restriction by the essentially two-dimensional nature of their living tissue and might thus be expected to continue growing at an undiminished rate for many years.

Growth measurements of single colonies over several years and comparisons of growth rates between colonies of different sizes reveal a pattern of accelerating growth in the first few years of life, when colonies are less than about 5 cm in diameter, followed by a fairly constant linear growth rate thereafter. A *Porites* sp. 5.8 m in diameter, estimated to be about 140 years old, has been found to increase in diameter at a rate of 4.3 cm a year, whilst excavations on the Pacific coast of Panama revealed branches of *Pocillopora* that must have grown continuously over several centuries. The potential longevity of corals is, however, unknown and it may be that most corals are killed long before reaching their maximum life-span, leaving no opportunity for us to witness growth rates in the senescent phase.

However, mechanical constraints will of course limit the maximum size attainable by branching corals. As these become bigger, branches become more liable to be snapped off by wave action and increasing load is placed on the relatively small area attached to the substratum, so the colony becomes increasingly unstable. The opposite is true of massive, especially spherical, 'brain' corals, which increase in mechanical stability as they get bigger. It does not follow from this alone that branching corals are likely to have shorter lives than massive corals. Broken fragments may survive and grow into large colonies, but these belong to the same clone as the original fragmented colony and constitute spatially separated parts of the same animal.

Other factors, however, may shorten the lives of branching corals relative to massive ones. Branching corals grow faster and presumably have a faster physiological turnover which, as in other animals, may signify a shorter life span. The small branching corals *Pocillopora damicornis* and *Stylophora pistillata*, for example, are quick-growing, colonizing species of short-life expectancy, adapted to the repeated recolonization of Indo-Pacific reef flats after periodic devastation by tidal emersion or by storms. Their growth rates decelerate with increasing age and colony size, and this growth pattern might well be found in other corals if growth could be measured at all stages in their potential life span.

6.4 Competition between corals

Branching corals, by virtue of their upright faster growth, will gradually overtop slower-growing competitors unless interrupted somehow. Shaded from the light, the overgrown species will eventually be ousted and recruitment of new colonies will be prevented, leaving a pure stand of branching corals. From sequential photographs taken over several years, Connell (1973) observed the branched colonies of *Acropora* gradually extending over encrusting colonies of *Montipora* on the Great Barrier Reef. After some of the *Acropora* branches had been broken off in a hurricane, the underlying portions of the encrusting colonies were seen to be dead.

In other situations the continued growth of branching corals may bring about their own demise. Buttresses, or spurs, on many windward Caribbean reefs are formed by the vigorous seaward growth of the elk-horn coral, *Acropora palmata*, the colonies of which become overcrowded and die, eventually to be grown over by other species.

Extracoelenteric digestion: competitive hierarchies

In 1973 Judith Lang published an account of interspecific aggression in corals that explains how the coexistence of fast- and slow-growing species may be possible. Within 0.5 to 12 hours of placing two species of coral in contact with each other, mesenterial filaments were extruded orally and through temporary openings in the polyp walls (Fig. 6.11). Usually one species would be more aggressive than the other and the mesenterial filaments of the more aggressive species would penetrate the adjacent polyp walls of the subordinate species. Within 12 hours, the tissue of the subordinate species in contact with the aggressive mesenterial filaments would be completely digested away, exposing the underlying skeleton. Such an attack proved fatal to colonies less than about 3 cm in diameter, but larger subordinate colonies suffered only local loss of tissue.

Each coral characteristically only attacked certain species,

Fig. 6.11 Extracoelenteric digestion in Caribbean corals. The large polyp of *Mussa angulosa* (right) has extruded some of its mesenteric filaments onto the subordinate species *Agaricia grahamae* (left) causing necrosis of the underlying tissue (pale). (After Lang 1973.)

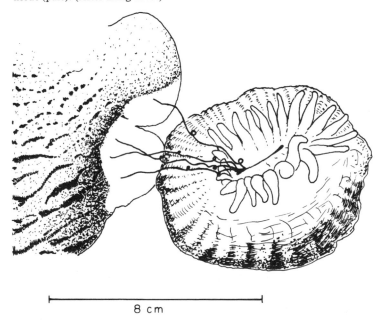

8 cm

being itself killed by certain other species, so that Lang was able to recognize an 'aggressive pecking order' among the Jamaican corals. The most aggressive species were slow-growing massive corals belonging to the families Mussidae, Meandrinidae and Faviidae. Intermediate in aggressiveness were fast-growing, branching acroporiid corals and least aggressive were foliose agariciids, also quick growers. Aggressors may attack more than one subordinate at a time, and corals of intermediate position in the aggressive hierarchy may attack a less aggressive coral even while being attacked on another side by a more aggressive coral. It appears that on Jamaican reefs, the aggressive hierarchy associated with extracoelenteric digestion promotes coexistence by allowing the slower-growing, massive species to protect themselves from overgrowth by the more rapidly expanding branching and foliose corals.

Sweeper tentacles: reversal of digestive dominance

This elegant tale of the counterbalancing processes of rapid growth and interference by extracoelenteric digestion may, however, take another turn on different reefs. Coral reefs on the Pacific coast of Panama are dominated in shallower water by species of *Pocillopora*, which are fast-growing, branching corals, and in deeper water by species of *Pavona*, which are slower-growing, massive corals. In the field, the distribution of scars left by extracoelenteric digestive encounters between neighbouring corals, suggests that *Pocillopora* is dominant over *Pavona*, opposite to what we should expect from the Jamaican situation. Yet in the laboratory, *Pavona* is able to damage the tissues of *Pocillopora* by mesenterial digestion within 12 hours of tissue contact.

Long-term experiments in the field have explained the paradox. After placing *Pocillopora* and *Pavona* together on the reef, within two days *Pavona* extends its mesenterial filaments and kills the adjacent tissues of *Pocillopora*. *Pavona* then retracts its mesenterial filaments and algae quickly cover the bared areas of *Pocillopora* skeleton. One to two months later, tissue regenerates over the bared patches, and the polyps on the peripheral branches of *Pocillopora* adjacent to *Pavona* convert some of their feeding tentacles into very elongated 'sweeper' tentacles that sway passively in the surge, frequently dragging over the *Pavona* colony. Contact by the sweeper tentacles causes necrosis and sloughing of the *Pavona* tissue. The exposed skeleton is rapidly colonized by filamentous algae and later by encrusting coralline algae that act as a buffer, preventing further tissue contact between the two corals. The sweeper tentacles of *Pocillopora* contract and resume their normal feeding function. Gradually, the faster-growing *Pocillopora* overtops the *Pavona*.

It is unclear why *Pavona* does not retaliate by extending its mesenterial filaments when under attack from the sweeper tentacles of *Pocillopora*. Perhaps the sweeper tentacles are more powerful than the mesenterial filaments. Sweeper tentacles were previously thought to be devices for intercepting zooplankton, but they are structurally similar to the special tentacles of sea anemones that are used for aggression between clones. In the feeding state, the short 1–2 mm tentacles of *Pocillopora* bear a higher proportion of spirocysts and a lower proportion of nematocysts than they do after elongation (30 mm), corresponding to a change in function between food capture and defence or aggression.

The significance of reversals in digestive dominance to the coexistence of fast- and slow-growing corals, remains to be fully worked out. Clearly the initial digestive dominance of massive corals, temporarily gained by the use of mesenterial filaments, cannot guarantee eventual coexistence. It will, however, delay the overtopping growth of faster-growing competitors. During this delay, the less robust branching corals may be damaged by predation or storms, so that mesenterial aggression coupled with the right timing of disturbance could save the slower-growing species from competitive exclusion.

Role of disturbance in preventing competitive exclusions

Any kind of disturbance that disrupts the process of competitive exclusion, but does not eliminate the competitors, will promote coexistence (p. 136). Disturbances on coral reefs are brought about either by biological factors, such as predation on the polyps or excavation of the coral skeleton, or by physical factors such as emersion on low spring tides or damage by storms. Whether or not the disturbances increase or decrease the number of coexisting coral species on a reef depends on how frequent, how extensive, how intense and how selective they are with regard to the species they affect.

Corals are not the only competitors for space on the reef, but must contend also with alcyonarions (soft corals) and algae. Proportions of these organisms recorded on a reef in the Red Sea were: on the upper fore-reef slope, corals 18–49%, soft corals 2–15%, algae 3–22%; and on the reef flat, corals 5–32%, soft corals 0.2–17%, algae 3–75%. The coexistence of corals, soft corals and algae depended on the renewal of empty space by various disturbances such as catastrophic low tides, predation and grazing. Sea-urchins and herbivorous fish prevented space monopolization by algae. Soft corals can be overtopped by corals but their rapid carpeting growth and apparent distastefulness to predators enable them to colonize rapidly any space reopened after a catastrophe.

6.5   Predation of corals

Predation of corals has hit the headlines in the past because of the spectacular inroads made by the crown-of-thorns starfish, *Acanthaster planci*, on certain Indo-Pacific reefs. Normally, *Acanthaster* occurs at fairly low densities and is an inconspicuous animal on the reef. In spite of its large size and heavy armoury of spines *Acanthaster* is often hard to see nestled in the coral. Some-times, however, *Acanthaster* forms large, startlingly dense aggregations (Fig. 6.12), and since the starfish feeds by everting its stomach and digesting the underlying coral tissue, such aggregations cause immense damage to the reefs, leaving behind large areas of white, denuded coral skeleton. Within two to three weeks the bare patches are colonized by filamentous algae, which

Fig. 6.12 Aggregation of *Acanthaster planci* feeding on *Montipora foliosa* on the Great Barrier Reef (viewed from above); white areas have been stripped of tissue. (After Endean 1973.)

are joined after a few months by encrusting calcareous algae and alcyonarians (soft corals). The more fragile skeletons of dead branching corals fragment into piles of rubble. New corals begin to recolonize the substratum within three to ten years, the speed of recolonization depending on the area damaged and the proximity of living corals as a source of planulae or living coral fragments. Recolonization is slow, partly because loose rubble and carpets of non-calcareous algae and alcyonarians hamper the settlement of planulae. Fast-growing, highly fecund 'pioneer' species of coral are the first to establish themselves and the recovery of the previous coral community structure may take 20–40 years.

When occurring at lower densities, *Acanthaster* does not have such a devastating effect on the reef and some have suggested that by preferentially preying upon competitively dominant corals, *Acanthaster* may actually promote coexistence of the competitors. Evidence from Panamanian Pacific reefs, however, seems contrary to this suggestion. Here *Acanthaster* normally ranges over the reef slopes and along the reef base where it feeds principally on the small colonies and broken living branches of *Pocillopora* spp. These potentially dominant species, however, have a size refuge from *Acanthaster*, partly because large colonies are less easily handled and partly because they harbour greater numbers of symbiotic crustaceans that repel the starfish.

The symbionts are a pistol-shrimp, *Alpheus lottini*, and a small crab, *Trapezia ferruginea*, both of which seek refuge among the branches of *Pocillopora* and feed upon the mucus liberated by the host. When *A. planci* attempts to mount *Pocillopora*, the pistol-shrimp waves its larger cheliped and snaps it vigorously, causing *Acanthaster* to withdraw quickly. The crab grasps the tube feet and ambulacral spines of *Acanthaster* and jerks them up and down, also causing the starfish to retreat. Predation by *Acanthaster* therefore reinforces the competitive dominance of branching over massive corals on the Panamanian reefs.

Other coral predators have been studied far less intensively, but the list of known coral-feeding animals includes fish, sea-urchins, crustaceans, polychaetes and gastropods. Some of these, such as small shrimps, crabs, and snails, specialize on a diet of coral mucus or tissue. Others, like the amphinomid polychaetes (so called fire worms because of their hollow, poison-filled bristles) include corals as only part of their diet. Fish, notably butterfly-fish, damselfish, wrasses, parrot-fish, surgeon-fish, trigger-fish, file-fish, puffer-fish, and porcupine fish, that have teeth and jaws modified for rasping and grazing, either feed directly on coral tissue or damage the skeleton accidentally while browsing for algae or hidden invertebrates.

Though the actual quantities of coral consumed by predators seem slight, the damaged tissue may not always be replaced and the bared skeleton may be invaded by other organisms that overgrow and eventually smother the colony, or that bore into the skeleton and weaken it. Important excavating organisms include filamentous green algae, sponges, polychaetes, barnacles, crabs, bivalves and gastropods. Some of these organisms can burrow directly into living corals (Fig. 6.13).

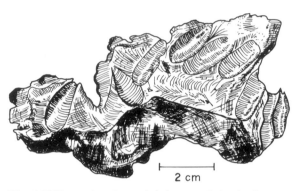

2 cm

Fig. 6.13 Excavations in coral skeleton made by the date mussel, *Lithophaga hanleyana*. (From Gohar & Soliman 1963.)

6.6 Environmental damage to corals

Tidal emersion

The principal environmental factors causing damage to corals are tidal emersion and storms. Only corals on the reef crest and reef flat are affected by tidal emersion, the amount of ensuing damage depending on the time of day and weather conditions that coincide with low tide. Early morning or evening low spring tides may cause no harm, but those occurring during the day may expose corals to intense insolation, damaging them by overheating, drying, or through the effect of light itself, which if too intense causes corals to extrude the zooxanthellae. Midday exposures have been seen to damage reef flat corals on the Great Barrier Reef, in the Red Sea and on the Pacific Coast of Panama. On the latter reefs, exceptionally low tides, − 70 cm below datum, exposed large tracts of *Pocillopora* spp. to the sun in January 1974. Hundreds of square metres of coral, whitened from the loss of tissue, were seen a few days after the exposure. Within two to three weeks algae began to grow over the dead coral skeletons.

Widespread damage to corals occurs when exposed by low tides to cold or rainy weather. Numerous colonies of *Acropora* on a reef flat on the Great Barrier Reef have been seen covered with a greyish fuzz of decomposing tissue several days after exceptionally cold weather coincided with low spring tides. *Acropora* colonies in a similar plight have been seen in the Seychelles after the corals had been exposed to prolonged heavy rain.

Quite often, it is only the upper portion of a coral colony that is killed by emersion. In the case of massive forms such as brain corals, repeated tidal emersion over the years leads to the formation of microatolls. Uppermost tissue is repeatedly killed so that the colony increases in girth but not in height. Erosion causes the upper surface of exposed skeleton to become slightly concave, giving the atoll shape.

Storm damage

Storm damage is perhaps the most significant kind of disturbance to corals down to moderate depths. Intermittent hurricanes and tropical storms affect coral reefs throughout much of their range, either passing close to the reefs themselves or generating heavy swell from a distance. The damage they cause depends on their intensity and distance from the reef.

Hurricane Gerta, with winds reaching 150 km h^{-1}, crossed the Belize barrier reef in the Caribbean in September 1978. Damage occurred all over the reef down to about 25 m. On the reef flat, corals were scoured of tissue by abrasive, suspended sediments and a number of large massive colonies were overturned. Below the reef crest, many corals on the spurs were broken off and deposited as loose colonies, broken branches and rubble in the grooves. In more protected areas, damage was very patchy: the complete demolition of stands of elk-horn coral occurred within a few metres of stands suffering minimal damage. Thickets of staghorn coral on the slope were flattened.

Colonies were seldom killed completely and both the dislodged fragments and remaining stumps retained patches of living tissue. Small fragments of broken coral were more abundant than large fragments, probably reflecting the greater fragility of branching corals and small, non-branching colonies. The percentage of fragments surviving increased with their size, exceeding 50% in those larger than 40 cm (Fig. 6.14).

Branches of the elk-horn coral, *Acropora palmata*, tend to

break off at the base, but because of their individual strength and high density they remain intact, lodged on the substratum close by, where they regenerate new colonies. The branches grow by about 7–8 cm a year and reach the 80% survival size in approximately ten years, which is the average interval between hurricanes at Belize. The periodic detachment and regeneration of branches may be an important mode of clonal growth and distribution for *Acropora palmata*, and perhaps also for other corals.

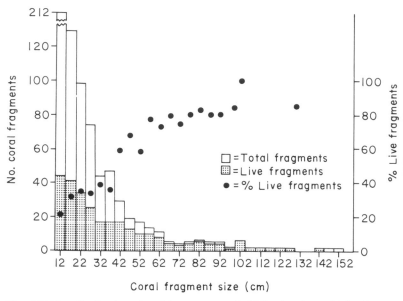

Fig. 6.14 Size distribution of coral fragments on the Belize barrier reef following Hurricane Gerta. (After Highsmith *et al.* 1980.)

Hurricane Gerta was of moderate intensity as far as hurricanes go. Very severe hurricanes can damage reefs so extensively, that instead of rejuvenating the reef by thinning out corals which have become crowded, there is almost total annihilation over wide areas. Hurricane Hattie passed across the barrier reef of British Honduras in 1961, destroying virtually all corals over a 43 km stretch of the reef and breaking up even the massive buttress formations (spurs) (see Fig. 6.7b). No regeneration was evident four years later and it was estimated that complete recovery would take 25–100 years.

6.7 Species diversity of associated coral-reef animals

By providing substrata for sedentary organisms and food or shelter for mobile animals, living bioherms and their geological end products create a rich series of habitats for an impressive diversity of species (p. 164). Is the diversity of these coral-reef inhabitants maintained by the same processes as that of the corals themselves? For most of the teeming invertebrates on reefs, the mechanisms of coexistence are completely unknown, but we have partial knowledge for two well-studied groups of animals: predatory gastropods and reef fish.

Predatory gastropods are particularly diverse on coral reefs, either living among the corals or confined to boulders, rubble or sediments on the reef flat. Why so many families, such as the Conidae, Muricidae, Terebridae and others, are better represented in the tropics than in temperate regions is partly a question of biogeography, to be considered in Chapter 10. But how can so many species of rather similar predators coexist on coral reefs?

Extensive work by Kohn on the genus *Conus* (Table 6.1) and other studies of the Muricidae and Terebridae have shown that each species tends to exploit a unique combination of available types of prey and microhabitats. Because of this 'niche specialization', interspecific competition is reduced and competitive exclusions, so common among corals competing for space and light, are probably rare. Unfortunately, it would be very difficult to test this inference by direct observation because of problems of sampling and of quantifying resource usage or of disentangling the effects of different population-controlling factors such as predation, competition and physical hazards.

Table 6.1 Major foods (percentages) of eight species of cone shells (*Conus*) on subtidal reefs in Hawaii. (After Kohn 1959.)

Species	Gatro-pods	Entero-pneusts	Nereids	Eunicea	Tere-bellids	Other polychaetes
flavidus		4			64	32
lividus	100	61		12	14	13
pennaceus	100					
abbreviatus				100		
ebraeus			15	82		3
sponsalis			46	50		4
rattus			23	77		
imperialis				27		73

Kohn (1979), however, made use of a 'natural' experiment by noting that *Conus miliaris* at Easter Island has a much broader diet than when it coexists with other congenors elsewhere in its geographical range. This supports the idea that on an evolutionary time scale, predatory gastropods respond to interspecific competition by niche contraction, thereby allowing coexistence.

A similar 'niche diversification' basis to coexistence would not be possible among the great number of coral species inhabiting a reef, because their essential resources of space and light could not be partitioned in so many ways as can the diverse prey and microhabitats used by gastropods. A certain amount of resource-partitioning is, of course, possible among corals with regard to autotrophy versus heterotrophy (p. 186), diurnal versus nocturnal feeding, depth, and exposure to wave action.

The species composition of certain reef fish assemblages requires yet another explanation. Damselfish (Pomacentridae) are small reef fish that use coral heads, or similar blocks of reef material, as a home base. Each coral head provides sufficient space for only a certain number of damselfish, which are territorial and repel intruders. By a long series of carefully replicated experiments and controls, in which the occupants of coral heads were manipulated, Sale (1980) has shown the local diversity of pomacentrids to be a function of the number of species available in the general area and of local population densities, with the result that the species composition at any site is unpredictable, depending on the history of chance colonization. There is no competitive hierarchy, as there is in corals, and no niche diversification, as there is in predatory gastropods. However, because of the limits to feasibility, Sale concentrated entirely upon a very small-scale habitat and a single guild of fish.

Longer-term studies of natural and artificial reefs in the Caribbean (Ogden & Ebersole 1981) suggest that although at the small scale of individual coral heads fish community structure may be a function of chance, at a larger scale encompassing the whole reef, fish community structure is a function of the orderly processes of competition and niche diversification. Communities of decapod crustaceans inhabiting coral heads in the Pacific seem also to be structured by chance events, and it would be interesting to see if evidence for more orderly processes would emerge from studies on a larger spatial scale.

6.8 Coevolution in coral reef communities

Disturbances and chance have been shown to be fundamental in the structuring of coral reef communities, and although these factors may explain how species can coexist on the reef, they do not explain why there are so many species in the first place. Probably there are several partial answers to this question, some of which will be considered in Chapter 10, but an important contributing factor is the overall predictability of the coral-reef environment.

Despite tectonic earth movements, eustatic changes in sea-level and changes in world climate, coral reefs rather like those existing today have persisted in the tropics for over 50 million years. Because corals require warm, clear waters throughout the year and are killed by relatively minor deviations from optimal conditions, the environment on reefs must have remained virtually constant throughout all those millions of years. This general stability over geological time has allowed ample opportunities for the coevolution of very specialized, finely tuned relationships between organisms and in this respect the old view of organismal diversity on coral reefs is correct.

The subtlety of coevolved biological interactions is epitomized in the bizarre symbiotic relationships to be found among coral reef inhabitants. Classical examples are the 'cleaner fish', usually small wrasses, butterfly-fish, or gobies, which feed on the ectoparasites attached to the outer surface, gills and buccal lining of larger fish (Fig. 6.15a). Parasite-laden fish queue up at certain 'cleaning stations', perhaps a conspicuous coral head, where they posture so as to allow the small cleaner fish access to the gills and mouth. The bold colour patterns and swerving, jerky movements of the cleaners (Fig. 6.15b) act as recognition stimuli to the hosts, thus ensuring that the cleaners are not eaten. A final twist to this coevolutionary tale is that cleaner-mimics occur (Fig. 6.15) which use the cover of similar recognition stimuli to gain close access to

Fig. 6.15 (a) Cleaner wrasse, *Labroides dimidiatus*, attending to the red snapper, *Lutianus sebae*. The mimic, *Aspidonotus taeniatus*, takes bites out of the 'host's' tail. (b) In addition to mimicking colour, *Aspidonotus* also mimics the undulating movement of *Labroides*. (After Wickler 1968.)

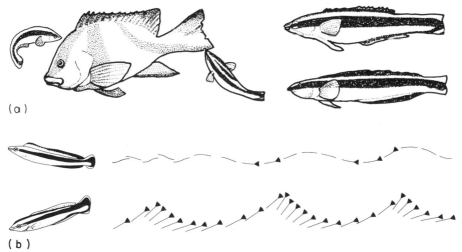

(a)

(b)

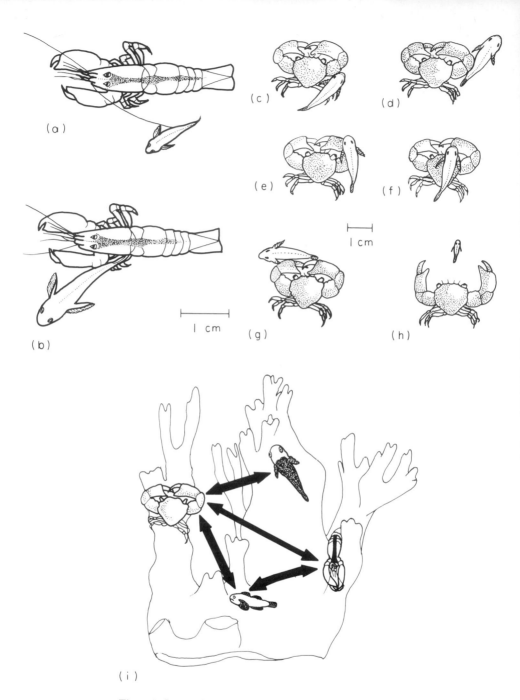

Fig. 6.16 Interactions between animals inhabiting the branched coral-head of *Pocillopora damicornis*. (a) Antennal contact between the shrimp, *Alpheus lottini*, and a juvenile fish, *Paragobiodon lacunicola*. (b) Body contact between a shrimp and an adult fish. (c–h) Spatial relations between the crab, *Trapezia cymodoce*, and the fish, *Paragobiodon* spp. (i) Summary of the interactions between the four species inhabiting *Pocillopora damicornis*: *Paragobiodon echinocephalus* (dark body), *P. lacunicola* (light body), *Trapezia cymodoce* and *Alpheus lottini*. The broad arrows indicate communication, incorporating the shivering behaviour of fish, the thin arrow represents the cleaning of the crab by the shrimp. (After Lassig 1977.)

the 'host', whereupon they take a bite of flesh rather than remove a parasite.

The intricate interdependence of organisms on coral reefs is not confined to symbioses. On the Great Barrier Reef for example, the branching coral *Pocillopora damicornis* harbours up to 16 species of crustaceans and fish that use the coral for food and shelter. Close study of the most conspicuous residents, two fish, a crab and a shrimp, showed that although they have the potential to harass, capture or exclude one another from the coral, a system of signals among the residents facilitates coexistence. The pistol-shrimp, *Alpheus lottini*, uses its legs to scratch the hairy hind surfaces of the claws of the small xanthid crab, *Trapezia cymodoce*. The crab performs similar movements itself to transfer food particles from the hairy margins of the claws to the mouth. By simulating this movement, the shrimp 'appeases' the crab. The two fish undergo shivering movements when in close contact with the shrimp or crab (Fig. 6.16), thereby preventing aggression from the crustaceans. Experiments showed that non-resident fish fled from approaching crabs, which became extremely aggressive and snapped at the fish. Newly colonizing fish have to out-manoeuvre the crustaceans among the coral branches until the signal system is learned. Once established, the signal system saves time and energy that would otherwise be spent evading the aggressive responses of the other residents. Each member of the established 'team' benefits by the presence of the others because the effort of defending the coral head against predators or competitors is shared. All inhabitants feed on coral mucus or tissue, and the repulsion of competitors may keep the number of residents at a level that is not detrimental to the food supply.

6.9   The trophic status of corals

Equally impressive as the species diversity of coral reefs is their high biological productivity, amounting to as much as 12 000 g C m^{-2} per year, which is maintained in seas that are otherwise relatively unproductive (20–40 g C m^{-2} per year) because of nutrient depletion (p. 6). The key to this high productivity lies in the symbiotic machinery of corals and zooxanthellae which facilitates a tight recycling of nutrients within the coral tissue. There is, however, much more to the story than that. Corals feed to varying extents on plankton, and perhaps also on bacteria that they trap in mucus or even on organic molecules taken up from solution across the body wall. These food sources contribute nutrient and energy to the total reef ecosystem, yet their relative importance in the nutrition of corals themselves remains in debate. A brief review of the known feeding biology of corals is therefore in order.

Autotrophs or heterotrophs?

In quick-growing, branching corals such as *Pocillopora damicornis*, the coral tissue forms a relatively thin layer over the skeleton, and the proportions of animal and plant protein in the tissue are about equal; in other words these corals are about half animal and half plant, whereas the proportion of plant material is less in the thicker tissues of massive corals. Evidently the branching corals function more like pure autotrophs and some of the massive corals more like pure heterotrophs. This divergent nutritional trend is reflected in both skeletal and polypary morphology. On passing through water, light is extensively scattered by suspended particles and molecules in solution. In shallow water the overall light intensity is so high that in addition to the vertical light from above, there is a considerable horizontal component of scattered light. The multilayered growth form of branching corals (see Fig. 6.2) increases the surface area of light-intercepting tissue both horizontally and vertically, enabling these corals to make maximum use of incident and scattered light. In addition to these skeletal modifications, the polyps of 'autotrophic' corals tend to be small, thereby exposing the maximum area of zooxanthellae, which reside mainly in the endoderm, to the light.

The more 'heterotrophic' corals typically have a monolayered, spheroidal skeletal structure (see Figs 6.2, 6.8). Normal plankton densities are such that after passing over a layer of filtering polyps and mucus traps, most of the potential food particles would be removed, so that the coral would gain very little from multiple layers of polyps. The tissues are thicker and the polyps larger in the 'heterotrophic' corals in accordance with their planktivorous habit. A spheroidal form maximizes the surface area of plankton-intercepting tissue, indeed as C. M. Yonge pointed out, corals have the highest prey-capturing surface to living-tissue volume of any animal. Encrusting and other growth forms are probably adaptations to non-trophic environmental factors.

Of course, light is important to all hermatypic corals, since without it they are unable to secrete large quantities of calcium carbonate, and probably all corals ingest some plankton; it is only the relative proportions in which these energy sources are used that differs among corals. Estimating these proportions is not easy and data are rather uncertain, but the proportion of energy ultimately derived from photosynthesis ranges from over 95% in the 'autotrophic' corals down to somewhat over 50% in the more extreme 'heterotrophic' species. To what extent corals acquire photosynthates by digesting the zooxanthellae or by taking up substances leaking across the algal cell walls is debatable, but the importance of the second mechanism is indicated by the ability of coral-tissue homogenates to stimulate zooxanthellae to release glycerol and small amounts of other organic molecules. Incubation of *Pocillopora damicornis* with $^{14}CO_2$ on the reef has shown that up to 50% of the ^{14}C fixed by the zooxanthellae during 24 h ends up in the coral tissue, mostly as lipid and protein. Evidently, the glycerol released by the algae is used by the animal in lipid synthesis.

Phosphorus and nitrogen may be limiting nutrients for algae in tropical seas, but zooxanthellae can obtain them from the metabolic wastes of their hosts, and owing to this recycling, the phosphorus and nitrogen turnover by corals is much slower per unit body weight than in other organisms (Fig. 6.17). Inorganic nutrients, however, must ultimately be taken up from sea-water by the corals. Two routes are possible: via consumed food particles or by active transport across the body wall.

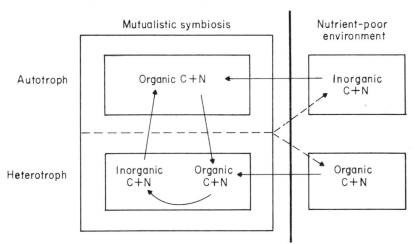

Fig. 6.17 Role of autotrophic symbionts within animal tissues in nutrient-poor environments.

Capture of food particles

Planktonic organisms are captured mainly by a mucociliary mechanism that also serves as a cleansing device, keeping the coral surface free of silt. Drifting particles that touch any part of the living surface of the coral become stuck in mucus and transported by ciliary currents either directly to the mouth or along rejection tracts if unsuitable as food. In some corals the food-laden mucus accumulates at the tips of the tentacles, which bend over and transfer the particles to the mouth, or bend outwards and reject them (Fig. 6.18). A correlation has been found between the amount of zooplankton material in the stomach and

Fig. 6.18 Methods of feeding in hermatypic corals. (a) *Euphyllia*, with large polyps with all cilia beating away from the mouth. (b) *Pocillopora*, with small polyps and cilia carrying particles up the column. (c) *Merulina*, a brain coral with short tentacles and reversal of cilia (the resultant current is indicated by the broken arrows). (d) *Coeloseris*, with all particles carried by cilia over the mouths for ingestion if edible and removal by water currents if not. (e) *Pachyseris* has mouths in rows within parallel grooves and no tentacles, but food is collected by mesenterial filaments extruded through the mouth. (After Yonge 1963.)

(a)   (b)

(c)   (d)

(e)

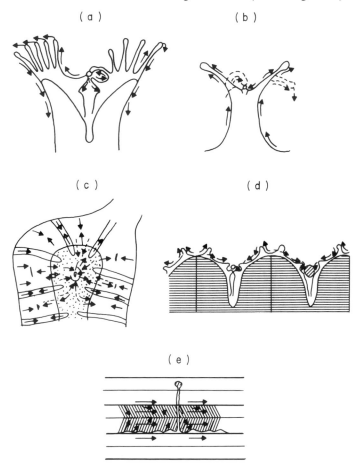

polyp diameter. Certain large-polyped corals are capable of using their long tentacles to sense microquantities of amino acids leaking from passing zooplankton and to capture the zooplankton passing within 1 cm of the coral head. Some corals use extracoelenteric digestion to cope with food particles too large to be handled either by the ciliary tracts or tentacles. Mesenterial filaments creep by ciliary action out through the mouth or through temporary openings in the body wall (see Fig. 6.11) on to the food material and proceed to digest it.

Bacteria rapidly colonize the mucus produced by corals as they do the mucus produced by any aquatic animal, but it appears that the amount of bacterial organic matter consumed by coral polyps may be very small under natural conditions.

Possible uptake of dissolved organic matter

Sea-water contains dissolved organic substances, ranging from free amino acids to macromolecules, which are a potential resource of those heterotrophs able to take them up against a concentration gradient (see p. 75). Some corals, notably those employing extracoelenteric digestion by the extrusion of mesenterial filaments, have an ectoderm covered with microvilli and possessing a high alkaline phosphatase activity. Such corals, therefore, seem well equipped, with an enormous absorptive surface and an enzyme system usually involved in membrane transport, for the active uptake of organic molecules. These features, together with the observed uptake of ^{14}C-labelled material, leave little doubt that corals can remove some substances from solution, but the extent of this remains to be established.

6.10   Productivity of coral reefs

Measuring the productivity of an entire coral reef is fraught with difficulties owing to the diversity of habitats from reef flat to reef slope and the diversity of organisms present in each of them. Added to these difficulties are uncertainties about the trophic status of many organisms, not least the corals themselves, and about the magnitudes of organic inputs to the ecosystem via oceanic plankton, bacteria and dissolved organic matter.

Primary production

Early attempts to estimate the overall primary productivity of reefs were based on day and night time comparisons of the change in oxygen tension that occurs in water as it flows over the reef. Such estimates indicated a high primary productivity of 1500–12 000 g C m^{-2} per year, which is two orders of magnitude greater than that of the surrounding seas. From more recent studies, a hazy picture of the relative contributions of benthic algae and symbiotic zooxanthellae to total primary production is beginning to emerge.

Benthic algae include the non-calcareous, frondose 'macro-algae' of inner reef flats, the belts of *Sargassum* present near the boulder zone on some reefs, upright calcareous algae such as the superabundant *Halimeda*, encrusting calcareous algae like *Porolithon*, filamentous algae growing either on surfaces or penetrating calcareous substrata such as coral skeletons, blue-green algae that grow as films on reef flats, and finally the zooxanthellae, present not only in corals but also in alcyonarians (soft corals), zooanthids (colonial anemones), giant clams (*Tridacna*) certain ascidians, and other animals.

After taking into account the proportions of total surface area covered by living *Acropora palmata*, macroalgae and filamentous turf-forming algae, together with the total surface area of substratum contained within a 1 m^2 quadrat, the net primary production of these autotrophic categories on a Caribbean reef has been estimated at 630 g C m^{-2} per year for the coral, 1170 g C m^{-2} per year for macroalgae, and 700 g C m^{-2} per year for the filamentous algae. The intrinsic net primary productivity of the zooxanthellae in *Acropora palmata* was about 10% that of the macroalgae and about 60% that of the filamentous algae.

Filamentous green algae are quick to colonize uncovered limestone surfaces, such as dead coral skeleton. They remain inconspicuous, however, because they are constantly grazed down by herbivores, such as guilds of browsing fish (p. 177) and gastropods, so that their standing biomass is negligible. Since the average herbivore biomass of some 150 g (dry weight) m^{-2} may be turned over at a rate of twelve times a year, the filamentous algae must be producing at least 3000–4000 g C m^{-2} a year, about five times that estimated directly. This discrepancy however, could easily be accounted for by uncertainties of the total area of filamentous green algae present on a reef. The encrusting calcareous alga *Porolithon* that dominates the algal ridges of Pacific windward reefs (p. 164), has an estimated net primary productivity of about 180 g C m^{-2} per year.

The overall picture of primary production is therefore very blurred. However, it appears that zooxanthellae in corals are about equally productive as the various kinds of benthic algae, but that because of their relative abundance compared with other autotrophs, zooxanthellae probably account for less than half the total net primary production in the coral reef ecosystem.

Adjacent to many reefs are large sea-grass beds covering sandy areas on the reef flat and producing around 7000 g C m^{-2} per year (Chapter 4). Only about 5% of this production is consumed by herbivores, the rest being degraded into detritus, some of which, together with the associated bacterial flora, will enter food chains on the adjacent reef. As the roots of sea-grasses fix nitrogen, they may be an important source of this otherwise scarce nutrient.

Atmospheric nitrogen is also fixed by blue-green algae, such as *Calothrix crustacea* that occurs on intertidal reef flats in the Pacific as a thin, monospecific film (it also occurs in other reef habitats in different growth forms). *Calothrix* can fix nitrogen at the rate of 1.8 kg ha^{-1} per day—two to five times the rate achieved by fields of lucerne or alfalfa. Fixed nitrogen enters the food web through at least three routes: (1) the blue-green algae are consumed by herbivores, especially by certain fish with low assimilation efficiencies, so that the water over the reef gains nitrogen via the fish faeces; (2) in areas subject to strong wave action, large pieces of the *Calothrix* film are dislodged and washed over the reef where they will be available to consumers; and (3) *Calothrix* releases about 50% of its fixed nitrogen into solution, from which it may be taken up by other autotrophs.

Nutrient input from oceanic plankton

The total autotrophic productivity of coral reefs is certainly very high, but does it balance the consumption rate of the heterotrophs? Some flow-respirometry studies show that it does, others that it does not. Nutrient cycling within the coral-reef ecosystem is highly efficient, at least 75% of the phosphorus pool of a reef is recycled within a day, but some losses are inevitable. To what extent is the reef ecosystem dependent on an external source of nutrient?

Plankton, as a source of energy and nutrients, is undoubtedly important on all reefs. Measurements of plankton depletion in water masses passing over a Caribbean reef have revealed a 91% reduction in the diatom crop and a 60% reduction in zooplankton, amounting to a total gain of about 65 g C m^{-2} yearly, equivalent to 4–13% of the net community metabolism and also 10% of the primary production.

191

Resident plankton

The figures above refer to net, incoming plankton to the reef, whereas up to 85% of the zooplankton ingested by corals may come from the reef itself. Resident zooplankton migrates vertically from under the coral heads at dusk and although eaten by the corals, this demersal plankton owes its existence to the shelter provided by the corals and associated reef structures, and represents an internal component of the coral-reef ecosystem distinct from the net input of plankton from beyond the reef.

The flow of energy between corals and resident zooplankton may be two-way. Corals produce about 50 mg of organic mucus m^{-3} per day in the Red Sea. Mucus contains energy-rich wax esters and so has a high total energy content of 22 J mg^{-1} (ash-free dry weight). The copepod *Acartia negligens* can assimilate at least 50% of the organic content of the mucus. Moreover, soon after its secretion as a clear, viscous liquid, mucus becomes denatured and transformed into polymeric strands and webs that tend to become suspended as flocs or incorporated into the detritus. The mucus soon gains a high bacterial content and there is little doubt that mucus and bacteria represent significant sources of energy in coral-reef food chains. Bacterial biomass has been estimated at about 60 mg C m^{-3} over the Great Barrier Reef compared with 20 mg C m^{-3} in the open water off the reef.

Conclusions

Despite all the uncertainties about the relative contributions to the coral-reef ecosystems of different kinds of autotroph, bacteria, dissolved organic matter, and internal versus external inputs, it is clear that the phenominally high total productivity is in large measure due to the combination of a tremendous surface area of photosynthetic tissue (either in the form of zooxanthellae or benthic algae and higher plants), optimal light and temperature conditions for photosynthesis, and the tight recycling of nutrients in an otherwise nutrient-poor environment. The efficient recycling of nutrients occurs both at the level of the coral–zooxanthella symbiosis and at the general level of the overall food web. Many consumers are present in reefs and although in absolute terms primary production is high, relative to the number of consumers, food can be regarded as scarce. Hence food is consumed rapidly and utilization is efficient. A high proportion of the environmental pool of nutrients is therefore maintained within living tissues, so reducing opportunities for the loss of nutrients out of the system. Any such losses are compensated by the slow accrual of nutrients from water masses passing over the reef and by the nitrogen-fixing activities of blue-green algal associations on the reef or rhizomes of adjacent sea-grasses. Summaries of recent work on coral reefs are to be found in Jones and Endean (1973) and Stoddart and Yonge (1971).

7 Pelagic and Benthic Systems of the Deep Sea

That the largest part of the ocean only merits the smallest chapter in this book is not a reflection of its relative importance but of our lack of knowledge. Nevertheless, we know very much more of deep-sea ecology now than we did twenty years ago (see Heezen & Hollister 1971; Menzies *et al.* 1973; Marshall 1979). In the early days of marine biology many people predicted that the abyssal depths would be found to be lifeless, whilst others thought (hoped?) that they would be populated by relics from bygone eras. We now know that it is neither so extreme nor a haven for competitively inferior lineages. Indeed, except for the absence of photosynthesis, the deep sea is not basically different from the regions described in Chapters 2 and 3, merely less productive and more sparsely inhabited as a consequence of the lower input of food materials. In this it is equivalent to cave ecosystems on land.

7.1 Pelagic zones

From the middle of the surface layer to about 2000 m, planktonic biomass declines almost exponentially and below that depth it

Fig. 7.1 Three profiles of zooplanktonic biomass in the Pacific Ocean. (From data in Menzies *et al.* 1973.)

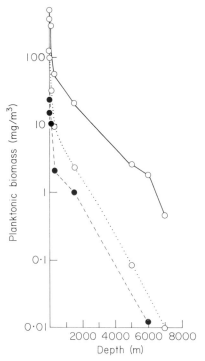

declines more slowly down to values of only a fraction of a milli-gram of living organisms per cubic metre of water (Fig. 7.1). Variation in planktonic abundance at a given depth correlates well with the productivity of the overlying surface waters, re-emphasizing that it is the rain of detrital particles and faecal material which supports the abyssal plankton. Most species are therefore omnivores or carnivores and these either seize macro-scopic items of food individually, or filter the microplanktonic consumers of bacteria. Vinogradov has argued for over twenty years that living food is introduced into the deep sea through an overlapping series of vertical migrations (Fig. 7.2) but although this is an attractive idea, there is very little evidence in support of it. Well-documented cases of vertical movements are all re-stricted to the upper 2000 m. The presence of algal debris in the guts of deep-sea species has been used in support of such a

Fig. 7.2 Schematic representation of a 'ladder of migration' in the deep sea. The variation in planktonic biomass with depth is also suggested by the frequency of dots in the left-hand column. (After Vinogradov 1961.)

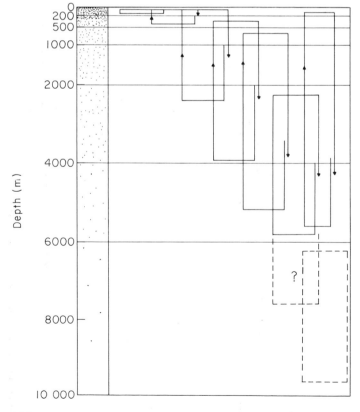

'ladder of migration' but these algae—especially the enigmatic 10–15 μm long 'olive-green cells'—are themselves inhabitants of the abyssal depths. Most probably the olive-green algae are the resting stages formed by diatom species as a reaction to sinking out of the photic zone: they have been recorded at all depths down to 5000 m. Populations of coccolithophores and blue-green algae of up to 350 000 individuals per litre are also known from depths of 1000–3000 m in the Atlantic and Mediterranean, although it is unlikely that they are active photosynthetically.

7.2 Benthic areas

Like the abyssal plankton, the biomass of the benthos declines markedly with depth, especially if the depth gradient coincides with increasing distance away from the coast (Table 7.1; Fig. 7.3). Most of the ocean bed supports less than 0.5 g (wet weight)

Table 7.1 Approximate average values of biomass (g (wet weight) m^{-2}) at different depths in the ocean.

Depth range (m)	Biomass
0– 200	200
500–1000	< 40
1000–1500	< 25
1500–2500	< 20
2500–4000	< 5
4000–5000	< 2
5000–7000	< 0.3
7000–9000	< 0.03
> 9000	< 0.01

Fig. 7.3 Variation in wet-weight biomass of the benthos of the Pacific Ocean in relation to depth and to distance from the nearest continental coast. (After Zenkevitch & Birstein 1956.)

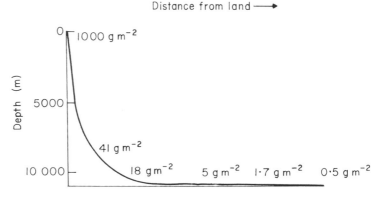

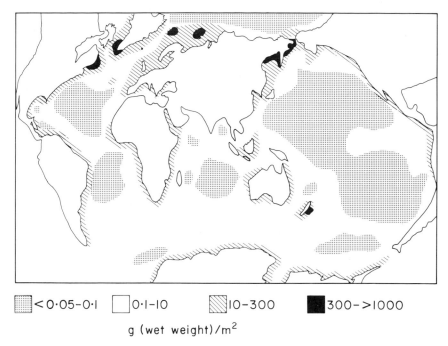

<0·05-0·1   0·1-10   10-300   300->1000

g (wet weight)/m^2

Fig. 7.4 The distribution of benthic biomass (wet weight) in the world ocean. (After Wolff 1977.)

m^{-2} of living organisms (Fig. 7.4). Because the deep-sea benthos is dependent on detritus and the associated bacteria, however, anomalously high biomass can occur where large quantities of detritus reach regions adjacent to a coast. Quantities of plant fragments and waterlogged wood have been found at 7000 m in the Banda Sea, for example, and in conjunction with this munificence were superabundant organisms—10–12 g (wet weight) m^{-2}—a very high biomass for that depth. Hydrothermal vents in the sea bed also support an anomalously large biomass, in this case dependent on the chemosynthetic sulphur bacteria which derive energy from the reduced sulphur compounds issuing from the fumaroles, and use it to fix carbon dioxide. Levels of organic carbon five hundred times the quantities normally to be expected are associated with fumaroles at 2500 m near the Galapagos Islands.

These, however, are exceptions: over most of the ocean bed the food input is very much less. Even in the relatively shallow depth of 2000 m, a mere 5.7 mg C m^{-2} per day was recorded by one study as sedimenting out of the water column, largely in the form of faecal pellets (660 pellets m^{-2} per day). Surprisingly large numbers of bacteria inhabit deep-sea sediments, however: one million bacteria on average per gram of sediment between 4000 and 10 000 m (up to a maximum of 84 million cm^{-3}), together with 20 000 protists cm^{-3} and 1.7–17 meiofaunal animals cm^{-3}. One might think that these densities would support considerable benthic productivity but it is very dangerous to use density or biomass as an indication of productivity, and in this case any impression of high potential productivity would be an illusion.

The metabolic rate of abyssal bacteria is up to one hundred times slower than that characterizing equivalent bacterial densities maintained in the dark, at the same temperature but at atmospheric pressure. This was discovered by accident. The Marine Biological Laboratory at Woods Hole, Massachusetts, operates a research submersible, the *Alvin*. In 1968, this was being prepared for a dive when it sank to a depth of 1500 m. The crew escaped but left their packed lunch (bologna sandwiches with mayonnaise, bouillon and apples) on the table and they left the escape hatch open. When the water-filled *Alvin* was recovered almost one year later, the lunch was found to be in perfect condition! This stimulated a series of experiments in which organic materials were maintained in parallel: (a) at deep-sea temperatures in the laboratory but at atmospheric pressure and (b) at 5300 m depth in the deep sea itself. These confirmed the very low rates of bacterial metabolism in the abyss. Bacterial productivities in the sea bed probably fall within the range 0.2 g C m^{-3} per day (at 1000 m) to 0.002 g C m^{-3} per day (at 5500 m). Consuming species must therefore be even less productive, relatively slow growing and long lived. The small bivalve *Tindaria* (8.5 mm in largest dimension) does not reproduce until it is at least 50 years old and can live to over 100 years. It therefore seems unlikely that the productivity of the deep-sea consumers exceeds 0.1 g C m^{-2} a year, except around fumaroles and on detrital bonanzas, and it must frequently be an order of magnitude less.

Although unproductive, the deep-sea benthos is diverse, especially on the continental slopes between depths of 2000 m and 3000 m. Even at 4700 m, single trawl hauls have yielded 196 macrofaunal species, but numbers of species per haul decline to twenty or less below 8000 m (Fig. 7.5). In part, the paucity of the fauna below 8000 m is an artifact of the sampling technique (trawls only operate well on level ground and cannot sample adequately the sides of the hadal trenches), but a number of animal

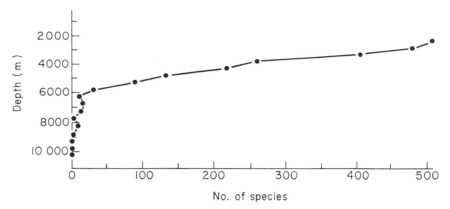

Fig. 7.5 The number of species of abyssal macrobenthos as a function of depth. (After Vinogradova 1962.)

groups do seem to have an apparent depth limit in the range 6200–8300 m. These include hydrozoans, flatworms, barnacles, cumacean, tanaid and mysid crustaceans, pycnogonids, gastropods, echinoid, ophiuroid and asteroid echinoderms, and fish; whilst decapod crustaceans only extend to some 5000 m. Many, though not all, of these absentees from the deepest regions are predators, and the groups which dominate the fauna below 8300 m are ooze-consumers such as the holothurians which comprise 90% of the faunal biomass in the deeper trenches (Fig. 7.6).

Whilst the meiofauna and the smaller members of the macrofauna (sometimes separated as the 'minifauna', macrofaunal species larger than 10 mm then forming the 'megafauna') are probably feeders on the bacteria and protists, the larger macrofauna are distributed between three or four feeding types: deposit (ooze) feeders, suspension feeders, and carnivores/scavengers. It is not possible, however, to assign species to these categories on the basis of the normal food materials taken by the groups to which they belong. Some deep-sea bryozoans and ascidians, for example, appear to feed on the deposits (whereas their shallower-water relatives are suspension feeders); and the giant, foraminiferan-like xenophyophorians of the abyss (some of which can achieve a diameter of 25 cm) may catch living members of the macrofauna. (In some areas, the pseudopodia of foraminiferans may carpet half of the available sediment surface.)

The relative importance of these feeding types is somewhat controversial. Recent Russian work has emphasized the importance of suspension feeders, especially in areas of the inert red clay ($< 0.25\%$ organic content), the organic poverty of which would be unlikely to support many deposit feeders. The Russian

199

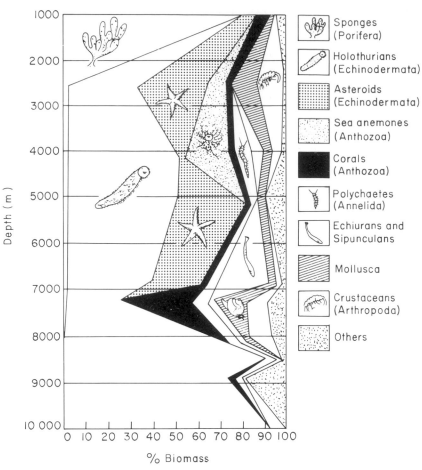

Fig. 7.6 The proportion of the benthic biomass at different depths comprised by different animal groups. (After Friedrich 1965.)

data, however, were gathered by the use of trawls, which might be expected to show a bias in favour of such suspension feeders as project above the sediment surface (sponges, crinoids, serpulid polychaetes, etc.). American work with box-corers, which take an *in situ* sample, has emphasized the importance of deposit-feeders, even in food-poor regions, but correspondingly this technique is biased against the larger suspension feeders. The two contrasting views are by no means mutually exclusive and it is probable that deposit feeders are the dominant component of the fauna in the more organically rich sediments (> 0.25% organic content) whilst suspension feeders, removing the sedimenting material from the water before it reaches the bottom, are most abundant on the hard substrata of the extensive mid-oceanic ridges, in areas of relatively rapid water flow (sediments may be ripple marked even at 7000 m), and in the relatively barren clays of the abyssal plain. The mobile and wide-ranging scavengers and carnivores are (not surprisingly considering their trophic position) less abundant than either of these two detritus-feeding types, but

200

when bait in the form of dead fish is lowered onto the sea bed, large numbers of ophiuroids, crustaceans and fish accumulate around it in a few hours. Such organisms have been termed 'croppers' and it is supposed that they roam over the deep-sea floor cropping any food organisms that they can find. Mobile deposit feeders (Fig. 7.7) may also act as croppers in that they will ingest other deposit feeders smaller than themselves.

Croppers are invoked by one faction in a current debate on the factors responsible for the high diversities of deep-sea faunas (see Chapter 10). One school of thought, based at the Scripps Institute of Oceanography in California, regards the feeding of indiscriminant predators as maintaining the populations of all the benthic prey species below the carrying capacity of their habitat (in terms of the quantities of bacteria/detrital-food available). Thus croppers prevent competition for food amongst the prey, competitive exclusion, and dominance by a few competitively superior species (p. 96). A second school, centred on the Woods Hole Laboratory on the other side of the USA, stresses the long-term implications of competition for food in the stable, presumed food-limited deep sea. Under this opposing hypothesis, competition for food has resulted in character displacement (p. 92), small niches, and the coexistence of many species as a result of

Fig. 7.7 Epifaunal holothurians (and brittle-stars) at a depth of 1060 m off California. (Official photograph US Navy.)

specialization on different parts of the total resource spectrum.
Much of the basic information which could resolve this debate
has not yet been gathered. We know nothing, for example, of the
extent to which the abyssal detritus feeders show dietary specialization: detritus feeders in general are often regarded as ingesting
material unselectively, but 'unexpected' selectivity has recently
been demonstrated in several intertidal species, and we are still
ignorant of what even the common littoral detritus feeders actually digest from their food in nature (see pp. 78–9). Neither do
we know to what extent the abyssal benthos is heavily predated.
Several aspects of their biology (e.g. slow rates of reproduction,
populations dominated by old individuals, etc.) do not suggest
that they have evolved strategies to cope with intense predation
pressure; and indeed we have seen above that predators are one
of the first feeding categories to disappear from the fauna with
increasing depth.

In any event, the two schools of thought are really considering
two different aspects of diversity. The Woods Hole school is postulating long-term mechanisms of diversity generation, whilst
that in the Scripps Institute is addressing the question of diversity maintenance at the present time. Over the deep sea as a

Table 7.2 Percentages of species present below 6000 m depth that are endemic
to the hadal region. (After data in Belyaev 1966.)

Group	No. hadal species	% endemic
Foraminifera	128	43
Porifera	26	88
Coelenterata	17	76
Polychaeta	42	52
Echiura	8	62
Sipuncula	4	0
Crustacea		
barnacles	3	33
cumaceans	9	100
tanaids	19	79
isopods	68	74
amphipods	18	83
Mollusca		
aplacophorans	3	0
gastropods	16	87
bivalves	39	85
Echinodermata		
crinoids	11	91
holothurians	28	68
starfish	14	57
brittlestars	6	67
Pogonophora	26	85
Vertebrata	4	75

Fig. 7.8 Giant, tubicolous, vestimentiferan worms (up to 3 m long) found abundantly at a depth of 2500 m on the Galapagos Ridge. These worms lack guts but have symbiotic, chemosynthetic, sulphur-oxidizing bacteria. (Photograph copyright by Woods Hole Oceanographic Institute.)

whole, trenches are clearly centres of the generation of new species (Table 7.2), but trench faunas have probably been derived from those of the abyssal plain rather than vice versa and the pattern of speciation on the abyssal plain is still problematic (p. 270). In Huston's argument (p. 96), predation is but one of the factors involved in the maintenance of diversity: to obtain a complete picture therefore, we need data on the rates of both potential population growth and population reduction. Growth rates must be very low in the deep sea and this means that only slight degrees of disturbance will serve to oppose competitive exclusion and maintain relatively high diversity, i.e. there is no need to postulate heavy cropping rates. Other probable disturbing factors are bioturbation, interference (pp. 84–7) and turbidity currents.

Clearly much more remains to be learned of the deep sea, including which organisms comprise its biota. In the last few years, several entirely new groups have been discovered (the vestimentiferan 'worms' (Fig. 7.8) and the xenophyophorian and komokiacean protists), and the faunal richness of some of the deep trenches is only beginning to be appreciated. Some trenches support 100 macrofaunal species m^{-1}. Photographs taken of the sea bed are also disclosing life-styles, patterns of distribution and abundance, and types of organism unknown from core and trawl samples (see Wolff 1977).

8 Fish and Other Nekton

All marine food webs, whether pelagic, benthic or fringing, terminate in the nekton: they comprise the upper portions of marine pyramids of consumption and production. Rather than treat them piecemeal in each of the previous six chapters, it makes greater ecological sense to accord these top consumers a section to themselves. Man, of course, sits on the pinnacle of the pyramid and consumes or otherwise utilizes fish, turtle, seal and whale products—although reciprocal consumption of man by nektonic animals is not infrequent—and this exploitation of nektonic production will be considered in Chapter 12. The present chapter will also introduce some of the basic ecological information upon which attempts to exploit the nekton 'rationally' are based.

8.1 Movements of the nekton

One of the features differentiating the nekton from the plankton is their ability to move in a horizontal plane relative to their inhabited water mass. Many species (and almost all of those of commercial interest) use this ability to maintain themselves in food-rich surroundings and to select habitat types which are optimal for the different stages in their life history. This is especially important where primary (and hence zooplanktonic) production is seasonal, i.e. in non-tropical latitudes.

In general, it is the young stages of most organisms which suffer the highest rates of mortality (see e.g. Table 8.1) and in

Table 8.1 Mortality rates for plaice, *Pleuronectes platessa*, of different age classes.

Age group	Mortality percentage per month
0–3 weeks	96
1–2 months	80
4–8 months	40
9–12 months	10
5–15 years	1

large measure this results from their relatively small size. Small organisms can be consumed by many more predatory species than can large ones, and hence it is selectively advantageous to grow through this critical stage as quickly as possible in order to achieve a size beyond the catching range of the abundant, small, sympatric predators. This can only be achieved in areas of relative food abundance (see also p. 82). For neritic/shelf species, one such series of areas-of-plenty is provided by regions of the littoral zone with extensive meadows of sea-grasses or with abundant populations of benthic algae, etc. Thus, semi-enclosed bays, lagoons and estuaries are often used as nursery areas by nektonic crustaceans, fish and even by the grey whale, *Eschrichtius gibbosus* (see Barnes 1974, 1980), and the littoral zone as a whole often supports high densities of larval and young nekton feeding on the abundant plankton and benthos. Such regions may not provide suitable food for the larger, adult individuals, however, and, accordingly, after growth to a certain threshold size, nektonic species may migrate offshore into deeper waters (Fig. 8.1).

Purely as a consequence of their size, the youngest stages of most nekton are planktonic and hence they are moved willy-nilly by surface currents. This constitutes a hazard with respect to successful arrival at the nursery grounds. Were the reproductive adults to spawn over the nursery ground itself, the larval stages might be carried out of many such areas by water flow, and, therefore, not only be unable to utilize their high productivity but also be carried into potentially unsuitable regions. Hence the adults must spawn at a site that will permit the drifting larvae to gain access to the nursery area at such a time as they are able to maintain themselves on station.

Fig. 8.1 The arrival, growth and departure of young flounder (*Platichthys flesus*) in a shallow lagoon in Denmark. (After Muus 1967.)

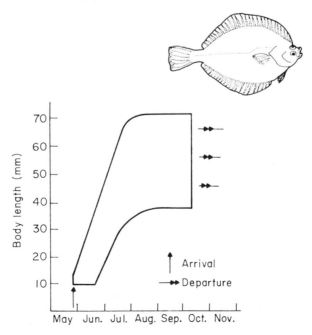

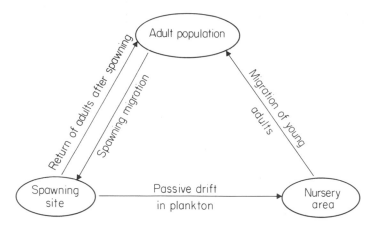

Fig. 8.2 The migration circuit of migratory nekton.

There are therefore three important points on the migration circuits of most species—spawning site, nursery area, and adult feeding grounds—and the whole circuit can be portrayed as a triangle (Fig. 8.2 and Harden Jones 1980), although it need not be triangular in geographical terms. In some deep-water species, the triangular form may be manifested in the vertical plane, with the eggs rising to the surface waters (or the adults migrating to spawn nearer the surface), larval life being passed in the photic zone, and the young adults migrating down to the aphotic depths. In some lagoons with minimal exchange of water with the adjacent sea, adult nekton may be able to spawn in the nursery area; and the same is true for those species which produce

Fig. 8.3 The probable migration route of the grey whale, *Eschrichtius gibbosus*, which feeds in the Chukchi Sea and calves in the bays and lagoons of the Gulf of California. (After Pike 1962.)

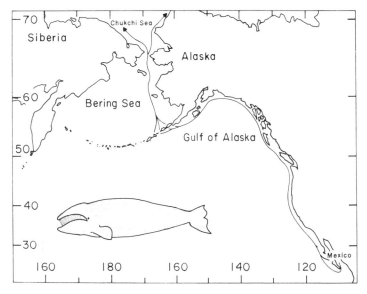

206

large young (e.g. the nektonic mammals), the migration circuit here being linear (Fig. 8.3). The basic triangle of migration, however, is shown by a wide variety of nektonic species; for example by shrimps, squid and fish (Figs. 8.4 and 8.5).

Large distances may be covered. The grey whale travels a distance of 18 000 km in each circuit and even the 30 cm-long herring may undertake a round trip of 3000 km. Although capable of swimming against currents, the nekton rarely undertake migrations in the face of opposing current flow, rather they use existing current patterns to minimize the energetic cost of travel. By regulating their depth in the water, they can make use of water masses at different depths flowing in different directions. Thus hake in the North Pacific may migrate southwards to spawn in the California Current, but return to the north in the deeper counter current. A number of fish species have also been shown to use tidal currents to aid their migrations—by resting on the sea bed when water flow is in the wrong direction but swimming up into mid water when currents are moving in the required direction of travel. The timing of the different phases of migration and the routes used appear stable over long periods of time (facts known and exploited by the fishing industry); the mean date of spawning in the Arcto–Norwegian cod stock, for example, has varied by less than ten days over the last 70 years.

Other species in seasonally productive waters make less

Fig. 8.4 Movements of the squid, *Todarodes pacificus*, in the Sea of Japan. (After Harden Jones 1980.)

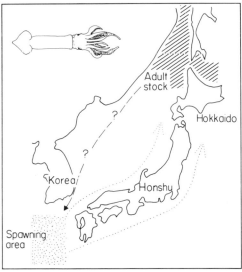

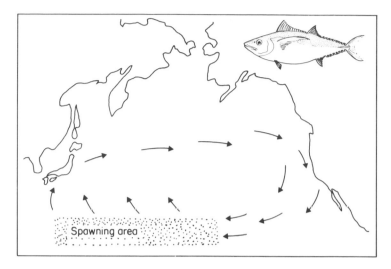

Fig. 8.5 Movements of the albacore, *Thunnus alalunga*, in the North Pacific Ocean. (After Harden Jones 1980.)

spectacular migrations to keep within areas of high productivity. Species exploiting the phytoplankton outburst of the Arctic summer, for example, move southwards during the autumn, spawn in lower latitudes and then follow the spring algal bloom as it spreads northwards. Nevertheless, the majority of nektonic species, especially those living within the shallow tropical regions of relatively constant productivity, do not migrate to any appreciable extent. They may show inshore movements to spawn, but their powers of locomotion are used to maintain station against water flow or to seek out concentrations or individual items of prey. Many species school, which may aid locomotion, the finding of prey, and anti-predator defence, and an increasing number are being found to use sound production as a means of communication and navigation.

8.2 Population dynamics

The dynamics of commercially important nektonic species are better known and have been more thoroughly investigated than those of any other marine organisms. The available information, however, is therefore biased in favour of migratory species (p. 204). (See Cushing (1975) and Gulland (1977) for reviews.)

The greatest hazards in the life of migratory—and many other—species occur when the individuals are youngest; in migratory species, between the spawning and nursery areas. One can assume that a given spawning site is used because it gives the greatest chance of successful entry to the nursery area. But it is optimally, not infallibly sited. If, by way of example, we consider the plaice, *Pleuronectes platessa*, in the southern North Sea (Fig. 8.6), it can be seen that successful establishment in the Waddenzee nursery will only result if several factors are all favourable. This population of plaice spawns over a period of two to three months, but with a peak on January 19th plus or minus less than a week. The eggs hatch after some fifteen days and larval life lasts an average of five weeks. During this time the young plaice are drifting as part of the plankton and, if all conditions have been favourable, they will metamorphose from pelagic larvae to benthic, postage stamp-sized versions of the adult fish, just as they arrive off the island of Texel. At some states of the tide, water moves from the North Sea into the Waddenzee through the 'Texel gate' and as the young fish descend towards the sea bed they are carried into their nursery ground, in which they will spend the first year or more of their life and grow from 1.5 cm to 20 cm in length.

Fig. 8.6 The migration circuit of the plaice, *Pleuronectes platessa*, in the North Sea.

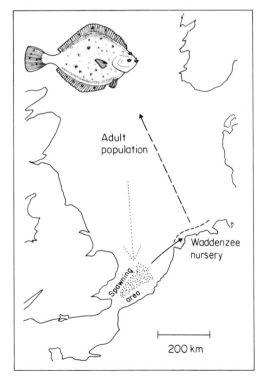

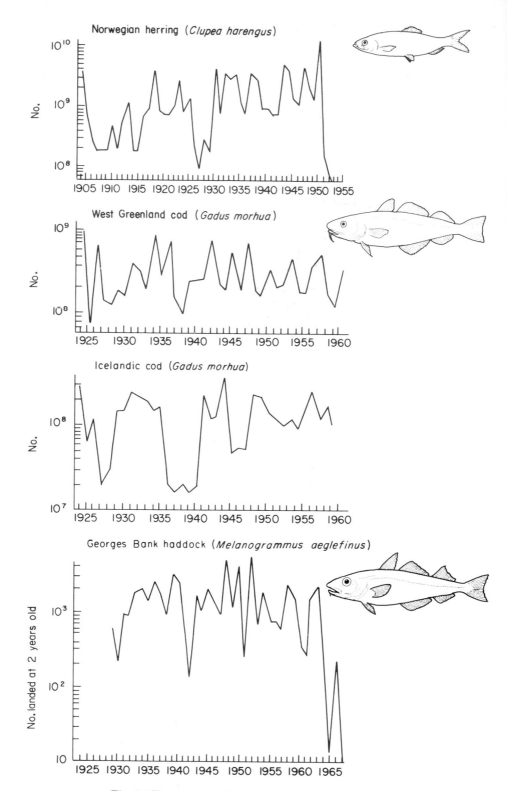

Fig. 8.7 Fluctuations in the strength of different year-classes of some commercial fish stocks. (After Cushing 1973.)

210

Current speed and direction are variable, however, and if the speed of drift is slower than average, the larval fish may metamorphose before they have arrived at the Texel gate. Alternatively, if the drift has been relatively fast, they may pass the critical entrance or pass out of the Waddenzee again whilst still planktonic. The timing of the spring phytoplankton bloom is also variable and so, therefore, will be the period of peak zooplankton abundance (upon which the larvae feed). If the bloom is late, the fish may starve before they can arrive off the Netherlands coast; if it is early, populations of potential competitors and predators may have already built up, resulting in particularly heavy mortality. Only about one year in seven can be classed as a favourable one. Plaice can spawn when four to five years old and can live to forty years old or more, although their life expectation once mature is only for a further nine years of life. On average, therefore, each adult plaice will only survive through less than two years in which larval survival is favourable. Mortality rates are high even in good years (Table 8.1) and so plaice produce some 500 000 eggs per year. Yearly fluctuations in the strength of recruitment to fish populations are by no means confined to the plaice (Fig. 8.7) and this can lead to populations being dominated by one particular year-class for many years (Fig. 8.8).

If a female fish during her lifetime produces five million eggs and the population is fairly constant in size, then there will clearly be a mortality of 99.99996% between egg and mature adult. If the adult population was halved, it would only require a reduction in this mortality to 99.99992% to return the adult population to its previous size. If fish compete with each other intraspecifically for food, then this small reduction in overall mortality might seem a feasible response to a halving of the original numbers of fish and the consequent reduction in competition intensity. Whether or not a relationship exists between size of the adult population and the numbers of young fish being recruited to it is obviously of great importance to fisheries scientists, as it will have a marked bearing on the ability of a given population to withstand an additional source of mortality. One problem associated with attempts to determine whether such a relationship does occur is the very wide variation in strength of year-class referred to above, resulting largely from factors other than intraspecific competition. In any event, in plaice for example, a mortality of 99.995% occurs in the first year of life whilst the fish are still larval or postlarval. Nevertheless, in some species, plots of recruitment (as measured by the numbers of young fish which have survived for long enough to enter the adult population and be

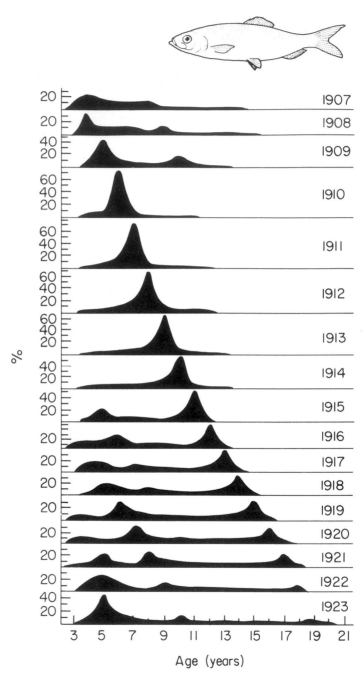

Fig. 8.8 Domination of a herring (*Clupea harengus*) population—as sampled by the Norwegian fishery—by the year-class which first entered the fishery in 1908 when four years old. (After Hardy 1959.)

caught by fisheries) against an estimate of the density of the total adult population do show curvilinear form (Fig. 8.9), suggesting an important density-dependent component of mortality. Such data are available for relatively few species, however, and in a number of these no good evidence for competition-induced mortality has been obtained. As theory would predict, those species which do show a relationship between these two variables have proved most able to persist in the face of human exploitation.

It is necessary here to add a brief comment on the assessment of nektonic population density. Commercial fisheries provide regular series of 'samples' of nektonic abundance but these data cannot be used directly as an estimate of population densities. Unit catches may have been obtained from vastly different volumes of water and may represent small samples from large, dense aggregations or the whole of a smaller, more diffuse population. Neither is weight of catch by itself a good measure: unit weight could comprise many small or fewer large individuals of a given species. With allowance made for the size and effectiveness of the nets used, for the extent to which fishing was only attempted in places or at times when fish were particularly abundant, and for

Fig. 8.9 Curve (with 95% confidence limits) relating the recruitment of different year-classes to a fish stock (R) to the numbers in that stock (P), both measured as numbers per unit area. Note the relationship between recruitment success and population density. Data are for the Arctic cod, *Gadus morhua*, stock. (After Cushing & Harris 1973.)

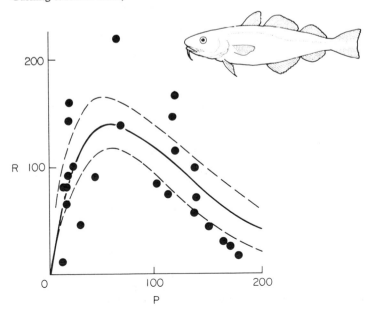

the size distribution of the catch, a good comparative estimate of abundance is the 'catch per unit effort'. This is only a relative measure, useful for demonstrating changes in population density with time, but it has proved capable of showing, for example, that although total catch may remain fairly constant, or even increase, the stocks being exploited can be declining rapidly (Fig. 8.10).

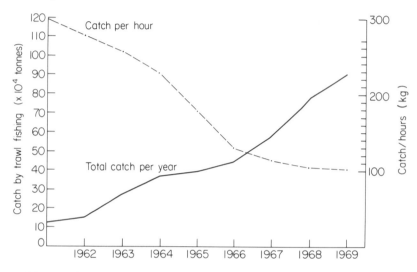

Fig. 8.10 Catch per unit effort and total catch by the Thai trawl fishery between 1961 and 1969. (From data in Shindo 1973.)

Not all nekton produce large numbers of young. The elasmobranch fish and marine mammals, in particular, bear few offspring; the larger rorquals, for example, probably can give birth to no more than eight young during their lives. In all such species, parental care is devoted to the embryos (which hatch/are born at an advanced stage of development) and this reduces the mortality experienced by other nekton when juvenile. A total mortality rate greater than 75% between birth and an age in the vicinity of thirty years will lead to a reduction in the populations of the larger whales: a reduction which human predation has demonstrably caused (pp. 297–9).

8.3 Nektonic production

The production of the nekton is secondary production and hence the discussion in section 2.4 is equally relevant here. Like that of the zooplankton, nektonic productivity broadly reflects the pattern of the primary fixation of carbon, but since most nektonic organisms are carnivorous, they are one or more stages further removed from the planktonic algae and littoral macrophytes. Indeed, energy fixed originally by photosynthesis may pass through very many organisms (both pelagic and benthic) before it is finally dissipated by the sharks and killer whales which are the top predators, and several of these intervening food species will themselves be nektonic. The productivity of the killer whales which feed on seals and of the sperm whales which take squid will therefore be very different from those of the herrings and anchovies which consume members of the plankton.

Ryther (1969) considered that, in general, there are fewer links between the nekton and the phytoplankton in upwelling zones than in oceanic waters (Fig. 8.11), mainly as a result of the relative size differences between the dominant phytoplankton and the nektonic consumers. If we use the average values of primary productivity given on pp. 37–9, Ryther's estimates of ecological efficiencies, and the numbers of intervening links in the food web displayed in Fig. 8.11, then the productivity of the planktivorous nekton and those species consuming other nektonic animals will be as set out in Table 8.2. Replacing Ryther's values of ecological efficiency with various others leads to wide variation in the estimates of nektonic production (Table 8.2); similar variation would also result from the adoption of alternative average numbers of links between the primary producers and the first nektonic

Fig. 8.11 Length of food chains in different marine systems.

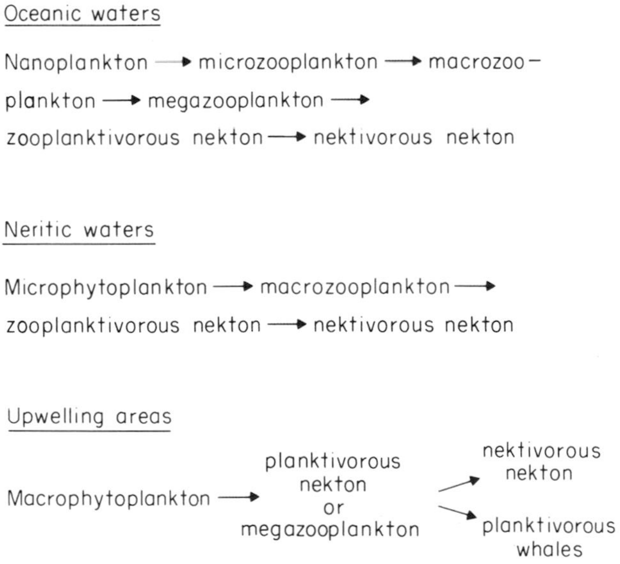

Oceanic waters

Nanoplankton ⟶ microzooplankton ⟶ macrozooplankton ⟶ megazooplankton ⟶
zooplanktivorous nekton ⟶ nektivorous nekton

Neritic waters

Microphytoplankton ⟶ macrozooplankton ⟶
zooplanktivorous nekton ⟶ nektivorous nekton

Upwelling areas

Macrophytoplankton ⟶ planktivorous nekton or megazooplankton ⟶ nektivorous nekton / planktivorous whales

Table 8.2 Four series of estimates of nektonic productivity all assuming: (a) net primary productivities of 50 g C m^{-2} per year (oceanic), 165 g C m^{-2} per year (neritic) and 230 g C m^{-2} per year (upwelling areas); and (b) food chains as shown in Fig. 8.11 (for upwelling areas, an average number of 1.5 links between the algae and the planktivorous nekton has been adopted). For each series of estimates, however, different values of the ecological or transfer efficiency between consumed and consumer are used. All estimates are in g C m^{-2} per year.

	Efficiencies after Ryther (1969)	Efficiencies after Cushing (1971)	Efficiencies after Ricker (1969)	Efficiencies as per p. 65
1 Oceanic nektonic productivity				
Planktivorous spp.	0.005	0.08	0.02	0.02
Nektivorous spp.	0.0005	0.02	0.002	0.002
2 Neritic nektonic productivity				
Planktivorous spp.	3.7	2.6	2.5	4.1
Nektivorous spp.	0.6	0.3	0.4	0.5
3 Nektonic productivity in upwelling areas				
Planktivorous spp.	20.6	2.6	8.9	16.3
Nektivorous spp.	4.1	0.1	1.3	2.0

consumers. These estimates suggest that the world production of planktivorous nekton lies within the range 74–152 $\times$ 10^{12} g C per year (the addition of benthic-feeding species and those consuming littoral vegetation would not alter these figures significantly). This is equivalent to 75–300 $\times$ 10^6 t (fresh weight) per year. By comparison, the total harvest by man of nektonic animals in 1977 was 57.5 $\times$ 10^6 t (fresh weight).

9 Ecology of Life Histories

9.1 Introduction

Amongst the bewildering diversity of marine organisms, run some basic patterns of life history that are correlated with the physical and biological nature of the habitat. The present chapter briefly reviews such patterns, interpreting wherever possible the significance of particular phenomena in terms of natural selection. In making these interpretations, it is necessary to keep in mind the counterbalancing benefits and costs associated with most adaptations. Energy or material committed to one function may be unavailable for another, and the development of a certain attribute may restrict the development of others. Each organism represents a compromise to the conflicting demands of simultaneous evolutionary problems. Excellence in one field usually costs poorer performance in another, with the result that different organisms outclass one another in different circumstances.

Early attempts to rationalize data on life histories often used MacArthur and Wilson's concept or r- and K-selection (e.g. Pianka 1974), where 'r' denotes the intrinsic (unlimited or exponential) per capita rate of increase and 'K' the carrying capacity (asymptotic population size) of the logistic growth equation. In the logistic model, growth rates of populations at densities well below the carrying capacity are influenced largely by the parameter r, while growth rates of populations at densities close to the carrying capacity are greatly influenced by K. Populations kept at, or repeatedly reduced to, low densities by physical or biotic environmental factors will be influenced by a different set of selective pressures (r-selection) from the set influencing more stable populations at high densities (K-selection). r-selected species correspond to opportunistic, early successional organisms in which rapid growth, early sexual maturity, high fecundity and great dispersal abilities are often advantageous. K-selected species correspond to late-successional organisms occupying more permanent habitats, in which competitive ability, resistance to predation, greater longevity, greater investment per offspring and the capacity to reproduce repeatedly in successive seasons are at a premium. r-selected features tend to be correlated with small body size and K-selected features with larger body size. As body size is inversely proportional to population growth rate and hence to the ratio of production to biomass, there is a trend of decreased population turnover rate and productivity from more r-selected to more K-selected species (Fig. 9.1.). The 'r–K continuum' has provided a framework upon which to build theories of life history, but the need for modification has become

217

increasingly apparent. Some of the organisms discussed below
(e.g. heteromorphic algae, benthic invertebrates with planktonic
larvae) will be seen to possess a mixture of *r*- and *K*- features,
while other properties, such as resistance to physiological stress
are not covered adequately by the '*r–K*' concept. The latter is
evidently too restrictive and is gradually being replaced by
more comprehensive schemes of classification as knowledge and
understanding of life history ecology develops.

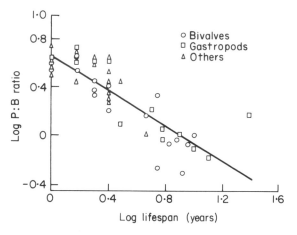

Fig. 9.1 The ratio of production to biomass decreases with increasing lifespan
and hence with increasing body size. (After Robertson 1979.)

9.2 Algae and higher plants

9.2.1. *Benthic forms*

Benthic vegetation extends to depths where the sea bed coincides
with the lower limit of the photic zone. Representatives from the
wide range of algal types, encompassing unicellular, filamentous,
encrusting and foliose forms, are potentially able to colonize any
kind of physical substratum. Due to their small size, rapid population growth and physiological robustness, unicellular and filamentous algae are able to survive under harsh or frequently disturbed conditions, such as the splash-zone of rocky shores, heavily grazed sublittoral rock surfaces or on the surfaces of mobile
sediments. In more equable circumstances, foliose and encrusting macroalgae establish themselves on hard substrata,
forming the dominant vegetation, e.g. intertidal fucoids or sublittoral kelps (Chapter 5).

Macroalgae are not so successful at colonizing sedimentary

substrata. Fucoids and kelps may be found attached to shells or small stones and may continue growing even when detached from the substratum in very sheltered localities. Indeed some members of the Fucales, such as certain morphs of *Ascophyllum nodosum* and the notorious oceanic rafts of *Sargassum* in tropical seas, can lead an entirely pelagic life. However, in general, angiosperms replace macroalgae as the dominant vegetation of shallow-water sediments, e.g. salt-marshes, sea-grass meadows, mangrove-swamps (Chapter 4). Features giving higher plants the ability to dominate terrestrial habitats also enable them to reinvade these aquatic habitats with equal success. Sophisticated skeletal and stomatal systems keep emergent vegetation erect and protected from desiccation while allowing efficient gaseous exchange. Roots and rhizomes provide anchorage, access to sedimentary nutrients and may serve as perennating organs protected from some of the physical hazards occurring above-ground. Angiosperm root-systems, however, are unable to cope effectively with hard surfaces, which therefore provide a competitive refuge for the macroalgae.

Macroalgae occur in an impressive variety of forms, but these can be grouped into several broad categories such as filamentous, membranous, finely branching, coarsely branching, calcareous and encrusting forms (Fig. 9.2a). Life cycles may involve quite different growth forms, e.g. minute gametophytic and large sporophytic phases of kelps (Fig. 5.13a), or the frondose gametophytic and encrusting sporophytic phases of certain red algae (for a review of red algal life cycles see Searles (1980)). Large physiological differences also occur among algae: some are quick growing but flimsy and short-lived, others slow growing, robust and long lived; some contain bacteriocidal and herbivore-repellent chemicals; some are resistant to considerable physiological stress and others less so.

These features represent counterbalancing capacities for rapid growth, reproduction, environmental tolerance, resistance to predation and competition for resources (nutrients, space, light). Among the possible combinations of such properties, certain groupings are particularly common, and many of these can be understood in terms of the type of habitat in which the algae are found. The role of physical and biological disturbances in determining community structure were discussed in sections 5.1, 5.2, and 6.4, from which it is evident that there exists a continuum from frequently disturbed habitats, such as heavily grazed, rocky surfaces, to relatively undisturbed habitats, such as the fucoid zone on sheltered shores. Combinations of features expected to

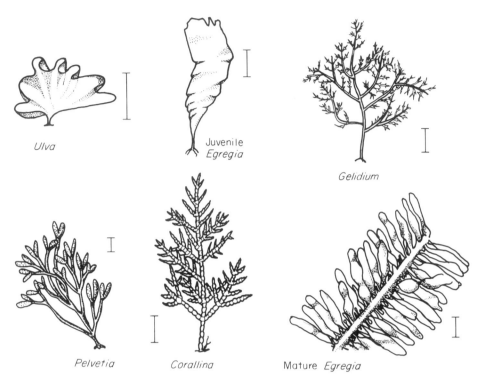

Fig. 9.2 A range of algal morphologies. Scales represent 2 cm. The segment of *Egregia* is from a frond about 4 m long. Facing: (b) Frond-toughness, percentage of structural (non-photosynthetic) material in the frond, and resistance to wave shock among a series of Californian intertidal algae. (c) Mean net primary productivities, energy contents, and palatabilities of the algal species as in (b). (After Littler & Littler 1980.)

be advantageous in species exploiting young, temporally fluctuating communities of disturbed habitats (corresponding to an *r*-selected regime, p. 217) are compared in Table 9.1 with those expected to be advantageous in species from mature, temporally constant communities from undisturbed habitats (similar to a *K*-selected regime). The counterbalancing benefits and costs ('trade-offs') associated with these features are listed in Table 9.2. To a large extent, theoretical expectations are born out in nature. Thus, membranous forms such as *Ulva* and *Porphyra* are commonly the earliest colonizers of disturbed, rocky intertidal substrata, and if succession is allowed to proceed, the early colonists are replaced by more robust species (Fig. 9.2). Opportunistic species have higher net productivities than later colonists, which invest more material and energy into non-photosynthesizing structural or defensive materials. This trend is reflected

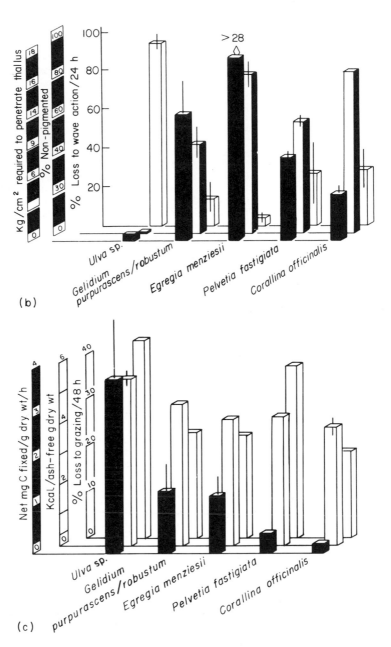

Fig. 9.2 cont.

(b)

(c)

by the ratio of productivity to total biomass (P/B), which decreases in the order of membranous, finely branching, coarsely
branching and encrusting forms. The high net productivities of
membranous algae such as *Ulva*, *Porphyra* and *Enteromorpha* are
partly attributable to the extremely thin construction of the thallus

Table 9.1 Hypothetical *a priori* survival strategies available to opportunistic macroalgae representative of stressed* communities versus macroalgae characteristic of nonstressed† communities. (After Littler & Littler 1980.)

Opportunistic forms	Late successional forms
1. Rapid colonizers on newly cleared surfaces.	1. Not rapid colonizers (present mostly in late seral stages) and invade pioneer communities on a predictable seasonal basis.
2. Ephemerals, annuals, or perennials with vegetative short-cuts to life history.	2. More complex and longer life histories; reproduction optimally timed seasonally.
3. Thallus form relatively simple (undifferentiated) and small with little biomass per thallus; high thallus area to volume ratio.	3. Thallus form differentiated structurally and functionally with much structural tissue (large thalli high in biomass); low thallus area to volume ratio.
4. Rapid growth potential and high net primary productivity per entire thallus; nearly all tissue is photosynthetic.	4. Slow growth and low net productivity per entire thallus unit due to respiration of non-photosynthetic tissue and reduced protoplasm per algal unit.
5. High total reproductive capacity with nearly all cells potentially reproductive and many reproductive bodies with little energy invested in each propagule; released throughout the year.	5. Low total reproductive capacity and specialized reproductive tissue with relatively high energy contained in individual propagules.
6. Calorific value high and uniform throughout the thallus.	6. Calorific value low in some structural components and distributed differentially in thallus parts. May store high-energy compounds for predictable harsh seasons.
7. Different parts of life history have similar opportunistic strategies; isomorphic alternation; young thalli just smaller versions of old.	7. Different parts of life history may have evolved markedly different strategies; heteromorphic alternation; young thalli may possess strategies paralleling opportunistic forms.
8. Escape predation by nature of their temporal and spatial unpredictability or by rapid growth (satiating herbivores).	8. Reduce palatability to predators by complex structural and chemical defenses.

* Young or temporally fluctuating.
† Mature, temporally constant.

and the large size of the cells, which result in relatively little self-shading by non-photosynthesizing materials. Late successional species grow more slowly and reproduce less profusely than opportunistic species, but their investment in non-photosynthesizing materials gains them greater life expectancy in at least three ways. First, non-photosynthesizing supporting tissue enables the algae to attain large sizes and to adopt

Table 9.2 Hypothetical costs and benefits of the survival strategies proposed in Table 9.1 for opportunistic (inconspicuous) and late successional (conspicuous) species of macroalgae. (After Littler & Littler 1980.)

Opportunistic forms	Late successional forms
Costs	
1. Reproductive bodies have a high mortality.	1. Slow growth, low net productivity per entire thallus unit results in long establishment times.
2. Small and simple thalli are easily outcompeted for light by tall canopy formers.	2. Low and infrequent output of reproductive bodies.
3. Delicate thalli are more easily crowded out and damaged by less delicate forms.	3. Low surface to volume ratios relatively ineffective for the uptake of low nutrient concentrations.
4. Thallus is relatively accessible and susceptible to grazing.	4. Overall mortality effects are more disastrous because of slow replacement times and overall lower densities.
5. Delicate thalli are easily torn away by the shearing forces of waves and abraded by the sedimentary particles.	5. Must commit a relatively large amount of energy and materials to protecting long-lived structures (energy that is thereby unavailable for growth and reproduction).
6. High surface to volume ratio results in greater desiccation when exposed to air.	6. Specialized physiologically and thus tend to be stenotopic.
7. Limited survival options due to less heterogeneity of life history phases.	7. Respiration costs high due to the maintenance of structural tissues (especially unfavourable growth conditions).
Benefits	
1. High productivity and rapid growth permits rapid invasion of primary substrates.	1. High quality of reproductive bodies (more energy per propagule) reduces mortality.
2. High and continuous output of reproductive bodies.	2. Differentiated structure (e.g. stipe) and large size increases competitive ability for light.
3. High surface to volume ratio favours rapid uptake of nutrients.	3. Structural specialization increases toughness and competitive ability for space.
4. Rapid replacement of tissues can minimize predation and overcome mortality effects.	4. Photosynthetic and reproductive structures are relatively inaccessible and resistant to grazing by epilithic herbivores.
5. Escape from predation by nature of their temporal and spatial unpredictability.	5. Resistant to physical stresses such as shearing and abrasion.
6. Not physiologically specialized and tend to be more eurytopic.	6. Low surface to volume ratio decreases water loss during exposure to air.
	7. More available survival options due to complex (heteromorphic) life-history strategies.
	8. Mechanisms for storing nutritive compounds, dropping costly parts, or shifting physiological patterns permit survival during unfavourable but predictable season.

overtopping growth forms that impart a competitive advantage over more delicate forms. The dominance of many sheltered, rocky shores by fucoids may be largely attributable to this factor. Second, the supporting tissue increases algal resistance to physical damage by waves, currents and scour. Third, a tough epidermis, calcareous matrix, low energy content, or presence of toxic chemicals reduce the acceptability to grazers.

Of course, organisms tend to be caught up in an evolutionary arms race in which defensive and offensive mechanisms are constantly readjusted with respect to each other. For example, calcareous matrices of encrusting *Lithothamnion*-like algae render these species unattractive to generalist herbivores such as periwinkles and many limpets. The South African territorial limpet, *Patella cochlear*, however, has acquired the capacity to deal effectively with *Lithothamnion*.

Tolerance to physiological stress, such as desiccation and extremes of temperature or salinity, may or may not be associated with reduced productivity. High-shore fucoids such as *Pelvetia canaliculata* are more resistant to desiccation, but are slower growing than lower-shore fucoids (p. 121). Some long-lived, slow-growing, encrusting algae such as *Hildenbrandia* are highly resistant to physiological stress (p. 123). On the other hand, opportunistic macroalgae such as *Ulva* and *Enteromorpha*, together with unicellular blue-green and diatomaceous algae, are not only fast growing, but also are among the most resistant algae to physiological stress.

A number of red algae form epilithic, encrusting sporophytes that are long-lived, apparently having envolved under selective pressures from intense grazing and stressful physical conditions, e.g. wave action, sand scour. They also form short-lived, upright gametophytes that are faster growing but less robust than the sporophytes. When first describing these 'heteromorphic' forms, taxonomists mistook them for separate species, and although their identity is now recognized, the nomenclature is retained. For example, a common heteromorphic red alga on western North American rocky shores is the encrusting sporophyte known as *Petrocelis middendorffii*, which alternates with the foliose gametophyte known as *Gigartina papillata*. The sporophyte grows slowly, increasing in area by about 4% a year, or even shrinking in some years, but is generally long-lived, an average-sized individual being anywhere from 25–90 years old. The gametophyte is probably annual, is more productive (0.64 versus 0.09 mg C g^{-1} per day) and is over twice as palatable to grazers as the sporophyte. Other algae, e.g. *Chondrus crispus*,

Corallina officianalis, may have extensive crustose basal or hold-fast systems from which grow erect thalli, and perhaps these algae invest differentially in the two structures according to environmental conditions. Among brown algae, kelps have heteromorphic life cycles (Fig. 6.13a), the tiny, ephemeral, opportunistic gametophytes alternating with the large, longer-lived sporophytes. However, even the sporophytes of some kelps may change morphologically as they grow. The young sporophytes of *Egregia* spp. (Fig. 6.13b) are thin, membranous forms with high growth rates enabling them to recruit effectively during early successional stages, but with increasing age the sporophytes invest more energy and material into supporting tissues, thereby reducing productivity but at the same time increasing competitive ability and physical robustness (Table 9.3). This shift in thallus form allows *Egregia* to compete effectively with opportunistic species for newly available space as well as to persist among late successional competitors for light and space.

Table 9.3 Comparative values for juvenile and mature individuals of *Egregia menziesii* used to test the shifting-strategy hypothesis. (After Littler & Littler 1980.)

Thallus used	Net productivity (mg C fixed g^{-1} (dry wt. per hour)	Toughness (kg cm^{-2} to penetrate thallus)	Time of appearance on successional plots (months)
Juvenile	$2.51 \pm .13$	$3.60 \pm .33$	3.0
Mature	$1.26 \pm .52$	> 28 (off scale)	10.0

The distribution of green, brown and red algae is very loosely related to depth. Green algae tend to be abundant intertidally, brown algae flourish both intertidally and sublittorally, while red algae are most numerous sublittorally. This distributional sequence used to be interpreted as chromatic adaptation to prevailing light conditions. Blue and green light is least absorbed by water, whereas longer wave lengths of red, orange and yellow light are strongly absorbed (see pp. 4, 43–44). Phycobilins and carotenoids of complementary colour to underwater light absorb the latter more efficiently than chlorophyll, on to which they are able to pass the excitation energy for photosynthesis. These accessory pigments give brown and red algae their characteristic colour, which has therefore been regarded as adapted to changing light quality with increasing depth. Accumulating documentation of distributional exceptions and conflicting physiological data, however, throw increasing doubt on the general applicability of the chromatic adaptation hypothesis. For example, green algae survive at the expense of brown and red algae when heavily grazed by sea-urchins in deep sublittoral areas (p. 146). Chromatic adaptation could still be important in allowing the red algae to outcompete the green algae when undisturbed, or to survive under the canopy of tall brown algae that also flourish at these depths, but its role is far from clear. Physiological experiments, moreover, have shown algal morphology to be at least as important as colour in determining potential vertical distribution (Ramus *et al.* 1976), whilst the actual limits to range are probably set by such factors as desiccation, grazing, competition and availability of sites (sections 5.1, 5.2).

9.2.2   *Phytoplankton*

Phytoplankton is the most abundant pelagic algal life and is virtually ubiquitous in the surface waters of the seas. Floating rafts of macroalgae, such as *Sargassum*, are very restricted in geographical distribution and are confined to the air-water interface. By existing as small suspended particles, phytoplankton gains access to subsurface nutrient supplies through transportation by vertical mixing. Small size is advantageous in a planktonic existence for reasons considered in Chapter 2.

Shape is also important: flattened discs, long cylinders or filaments sink slower than spheres of similar volume. Spheres and discs absorb nutrients faster per unit mass than long filaments of similar diameter when sinking at the same rate. Elaborations in the form of spines or hairs (Fig. 9.3) may reduce sinking rates in some instances, but because of the relatively high density of cell wall material, pronounced ornamentation could increase sinking rates in other cases. Large, spiny diatoms and dinoflagellates frequently contain oil droplets and large vacuoles that increase their buoyancy and it would appear that the function of spines and other processes is frequently to deter zooplanktonic grazers such as copepods. Some diatoms are enclosed in gelatinous capsules of only very slight excess density. The gelatinous capsules may protect the diatoms from predation either by making them too big for zooplanktonic grazers to handle or by allowing them to pass unharmed through the guts of their predators. Chain-formation may decrease or increase sinking rate according to whether the aggregative surface area to volume ratio is increased or decreased (Fig. 9.3) and may deter some predators. Size and shape may therefore represent an evolved compromise to the effects of

226

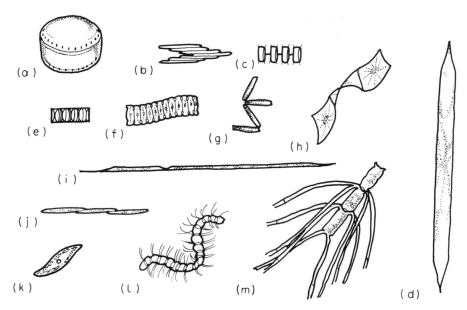

Fig. 9.3 Adaptive shapes and ornamentations of planktonic algae: (a)
Coscinodiscus concinnus; (b) *Bacillaria paradoxa*; (c) *Thalassiosira gravida*;
(d) *Rhizosolenia styliformis*; (e) *Paralia sulcata*; (f) *Bellarochea malleus*;
(g) *Thalassiothrix nitschioides*; (h) *Streptotheca thamensis*; (i) *Rhizosolenia hebetata*;
(j) *Nitschia seriata*; (k) *Gyrosigma* sp.; (l) *Chaetoceros curvisetus*; and
(m) *Chaetoceros convolutus*. (After Hardy 1962.)

sinking rate, efficiency of nutrient uptake, predation and possibly
other factors (Hutchinson 1967).

Although planktonic algae exist as single cells or small cellular
aggregates it is important to realize that the cells divide mitoti-
cally under favourable circumstances to produce clones of cells
with identical genotypes. Each clone is genetically equivalent to
the multicellular body of a non-clonal alga such as a fucoid or a
kelp, and in this sense the clone is the genetic 'individual' and the
component cells are spatially scattered modules of its 'soma' (for
a discussion of clonal population ecology see Harper (1977)).
Advantages gained by dividing the body into a large number of
detached modules include: (1) a high productivity facilitated by
high ratios of absorptive surface area to cytoplasmic volume and
of photosynthetic to structural materials; (2) the potential ability
to disperse modules among different water masses, thereby
increasing the chance of encountering favourable growing conditions; (3) the ability to quickly exploit localized resources by
rapid (exponential) clonal growth; and (4) lessening the probability of extinction of the genotype by spreading the risks of mortality among many independent modules.

9.3 Animals

Animals lead more varied lives than algae and higher plants, largely due to the more diverse sources of energy and nutrients that they exploit. The richness of life histories of marine animals could exceed the capacity of even a large textbook, but some important general features will be discussed here under the arbitrary headings of feeding and reproduction.

9.3.1 Feeding

9.3.1.1 Filter- and deposit-feeders

Just as benthic algae and higher plants can acquire energy by remaining stationary and intercepting sunlight, so numerous animals can meet their energy requirements by remaining attached to the substratum and intercepting water-born food particles: either planktonic organisms or particulate organic matter. Filter-feeders sift food particles still in suspension, while deposit-feeders gather food particles that have fallen out of suspension into the sediment. Filter-feeders are represented among diverse phyla (Fig. 9.4), being most numerous where currents bring food particles from large catchment areas. Many filter-feeders employ ciliary tracts, sticky mucus, or both, to trap food particles and transport them to the mouth. Arthropods lack cilia and use meshes constructed from interlocking hairs and bristles (Fig. 9.4g). The filtration device is either external, as with the radially symmetrical crown of tentacles or bristles of coelenterates, bryozoans, phoronids, polychaetes, arthropods, echinoderms and hemichordates, or internal as with the choanocyte chambers of sponges, lamellae of bivalves, lophophores of brachiopods and pharyngeal chambers of tunicates. Radially symmetrical discs, cups or fans formed by tentacles or intermeshing bristles, optimize combinations of surface area, mechanical strength, metabolic operational costs and trapping efficiency. The latter may also be increased by adjusting the orientation of the filter so that it always faces the water currents. This may involve movement of the animal, as in barnacles and polychaetes, or a modification of colonial growth as in feather- or leaf-shaped hydroids, bryozoans and gorgonians that grow with their broad surfaces at right angles to the current (Fig. 9.4c).

Coelenterates, through their possession of stinging nematocysts, are potentially able to catch large prey in addition to small suspended particles. Some anemones, such as the western

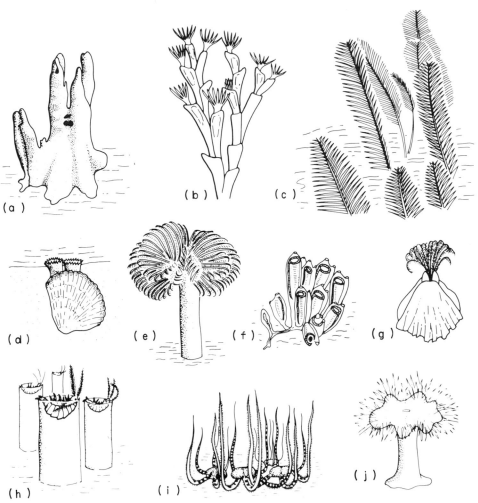

Fig. 9.4 Filter-feeding benthic animals: (a) the sponge *Amphilectus*; (b) the bryozoan *Bugula*; (c) the hydroid *Aglaophenia* with its flat surface at right angles to the prevailing current; (d) the bivalve *Cerastoderma edule*; (e) the polychaete *Bispira volutacornis*; (f) the tunicate *Clavellina lepadiformis*; (g) the barnacle *Balanus balanoides*; (h) the amphipod *Haploops tubicola*; (i) the brittle star *Ophiothrix fragilis*; and (j) the anemone *Metridium senile*. (After Hughes 1980b.)

North American *Anthopleura elegantissima*, that subsists largely on dislodged mussels, can cope with prey almost as large as themselves. Other filter-feeders, e.g. stony corals, soft corals, giant clams (*Tridacna*) and certain ascidians living in shallow, sunlit waters, rely to a greater or lesser extent on the photosynthates of symbiotic zooxanthellae (dinoflagellates, or in the case of tunicates, blue-green algae) as a source of energy (p. 190).

Deposit-feeders also comprise a diverse taxonomic assemblage (Fig. 9.5). Whereas filter-feeders tend to live on hard substrata or in sediments where there is little silt to clog the delicate filtering mechanisms, deposit-feeders are able to cope with silty sediments which they ingest to extract the micro-organisms associated with

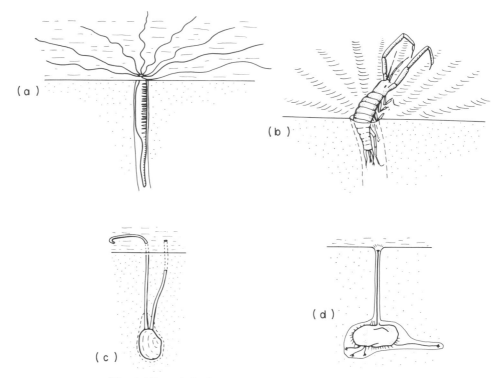

Fig. 9.5 Deposit-feeders: (a) the polychaete *Amphitrite*; (b) the amphipod *Corophium volutator*; (c) the bivalve *Scrobicularia plana*; and (d) the burrowing urchin *Echinocardium cordatum*. (After Hughes 1980b.)

detrital or mineral particles. Because deposit-feeders need to process copious quantities of sediment to extract sufficient food and because they also tend to live in unstable substrata, they are generally more mobile than filter-feeders, showing greater morphological adaptation for locomotion and lacking the radial symmetry and colonial existence of many filter-feeders.

Predators

In an ecological context, herbivory and carnivory can be regarded as types of predation, the only essential difference being the trophic level of the prey. Predators may consume entire prey organisms, as with fish eating zooplankton or dogwhelks eating barnacles, or they may consume only part of the prey organism without killing it, as with herbivorous grazers of benthic macroalgae or carnivorous grazers of sedentary colonial animals. Both grazers and non-grazers show a wide range of feeding behaviour from extreme dietary specialization to extreme generalism. Dietary specialization is more appropriate where specific prey are predictably abundant. Such is often the case with prey that have defence mechanisms effective against predators, resulting in low overall predation pressure. Examples are plentiful among sedentary animals. Sponges are pervaded by spicules that not only make sponge tissue unpalatable to many predators, but also lower its energetic value as food. In tropical to warm temperate regions where grazing fish are numerous, sponges living on open surfaces contain toxic chemicals that repel fish; for example, in the Red Sea, the sponge *Latrunculia magnifica* contains a cholinesterase inhibitor. In the same regions, sponges occupying protective microhabitats within crevices or beneath ledges, do not possess toxins, which presumably are manufactured at some cost. Coelenterates are charged with stinging nematocysts that are defensive as well as offensive in function. The sea anemone *Anemonia sulcata* also contains toxic polypeptides that are lethal when experimentally injected into fish and crustaceans, and which presumably repel predators in nature. Organisms protected from the majority of predators present a potentially rich food source for any animals that develop means of breaking through the defence mechanisms. Not surprisingly therefore, virtually all plants and animals armed with antipredation devices are exploited by a few coevolved predators, specialized to deal with these defences. Certain nudibranches and cowries specialize on sponges; for example the nudibranch *Archidoris pseudoarga* feeds entirely upon the sponge *Halichondria panacea*, and the nudibranch *Aeolidia pappilosa* feeds on sea anemones.

Generalist predators can be expected to contract or expand their diets according to the relative abundances of more or less preferable prey. The western North American starfish *Pisaster ochraceus* prefers mussels but will also feed on barnacles and then gastropods as the preferred prey become scarce. Dietary preferences sometimes reflect the profitabilities of different prey. Profitability will be a complex function of the likelihood of capture, yield of food material and the time and effort needed to handle the prey. In other instances, dietary preferences may reflect the limited time available for foraging or the need to minimize the time spent foraging, during which the predator may itself be at risk to predation or to other mortality factors.

Predators can be classified as searchers, pursuers and ambushers (Fig. 9.6). Searchers spend most of their foraging time locating prey, which once encountered are caught and eaten

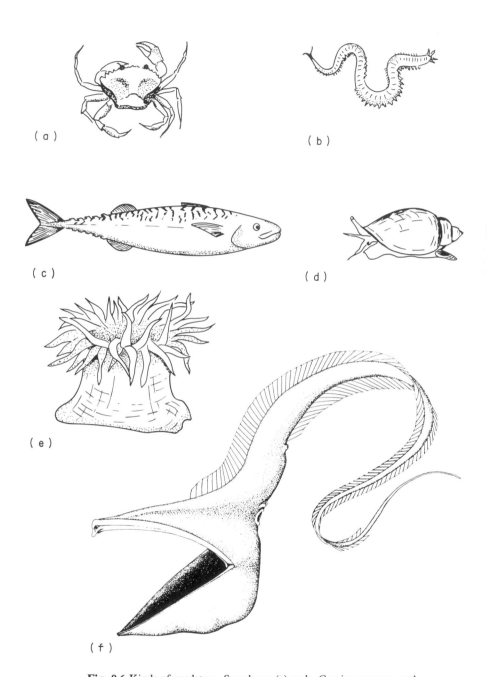

(a)

(b)

(c)

(d)

(e)

(f)

Fig. 9.6 Kinds of predators. Searchers: (a) crab, *Carcinus maenas*, and
(b) polychaete, *Nereis* sp. Pursuers: (c) mackerel, *Scomber scombrus*, and
(d) dogwhelk, *Thais lapillus*. Ambushers: (e) anemone, *Actinia equina*, and
(f) deep-sea gulper, *Eupharynx pelecanoides*. (a and c after Hughes 1980b; d and e
after Hughes 1980a; f after Briggs 1974).

relatively quickly; these predators will tend to be generalists, feeding opportunistically on any easily caught prey. Pursuers feed on prey that take a relatively long time to catch and subdue, so that a large proportion of the foraging time is spent pursuing individual prey; these predators will tend to specialize on prey with high profitabilities, perhaps becoming morphologically adapted to capturing particular prey efficiently. Ambushers wait for prey to come close enough to be caught by surprise or by blundering into a trapping mechanism; these predators do not actively search for prey or pursue them, but some considerable time may be spent subduing and digesting very large prey. Ambushers cannot determine which prey shall be encountered and so will tend to accept all capturable prey, sometimes being able to tackle prey as large as or even larger than themselves.

As with most biological classifications, this categorization of predators is arbitrary and many animals will have intermediate predatory behaviour or may even switch from one category to another according to circumstances. The shore crab, *Carcinus maenas*, is a most opportunistic feeder as it migrates with the tide to forage on the shore (Fig. 9.6a). When feeding among an abundance of mussels, however, *Carcinus* displays a surprising ability to modify its predation technique to deal specifically with mussels and tends to choose the most profitable mussel sizes (Fig. 9.7). The dogwhelk, *Thais lapillus*, feeds almost entirely on barnacles and small- to medium-sized mussels. Although these sedentary prey are not pursued, dogwhelks spend a very long time handling them (over 10 h to drill and over 15 h to consume a medium sized mussel) and because of this, dogwhelks fit into the 'pursuer' category. However, like most predators, dogwhelks will feed opportunistically on any moribund prey, such as a freshly killed fish. Ambushers are often morphologically constrained to adopt the one predation method. Anemones such as *Anthopleura elegantissima* or *Actinia equina* that feed on macroscopic food, are unable to pursue prey, but their stinging nematocysts and numerous tentacles enable them to subdue large prey which slip through the wide pharynx, lubricated with mucus, into the distensible sack-like stomach. Deep-sea fish such as gulper eels (Fig. 9.6f) and angler-fish are ambushers *par excellence*. At these great depths prey are scarce indeed (see p. 29), so that pursuit and selective feeding would be uneconomical. Gulper eels, angler-fish and others have become little more than suspended traps, most of the body muscles and skeleton having atrophied except for the jaw apparatus, so that sustained searching or pursuing would be impossible. The mouth has an enormous gape,

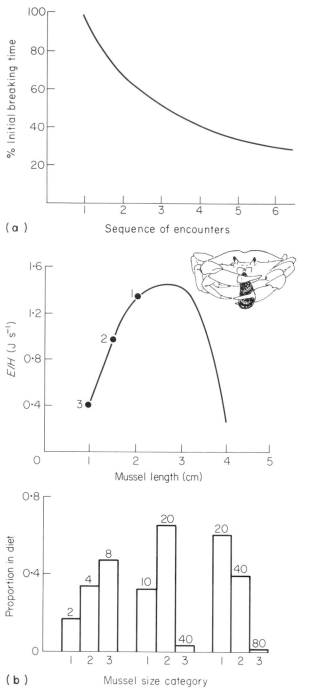

(a)

Sequence of encounters

Mussel length (cm)

(b)

Mussel size category

Fig. 9.7 (a) The time taken by the crab *Carcinus maenas* to open a mussel, *Mytilus edulis*, decreases as the crab becomes more experienced at handling the prey. (After Cunningham & Hughes in prep.) (Upper b) The profitability (energy yield per handling time) of mussels to foraging crabs peaks at intermediate mussel sizes. (Lower b) The diet of shore crabs fed on three sizes of mussels where the profitability of 1 > 2 > 3. Numbers offered appear over the histograms. As the most profitable mussels became more abundant the crabs fed disproportionately upon them. (After Townsend & Hughes 1981.)

and in some cases can even be unhinged, while the stomach is hugely distensible so that prey considerably larger than the predator can be swallowed.

Copepods, which account for a large proportion of the zooplankton, are important filter-feeders on phytoplankton (Chapter 2) and as with all arthropods, the filter is constructed from intermeshing bristles born on modified limbs. Size selection of algal cells occurs as a mechanical consequence of the mesh diameter of the filter, but whether mesh size can be modified according to the availability of different algal types remains in debate. Algal quality also is important in addition to size. *Acartia clausi* avoids the dinoflagellate *Ceratium tripos*, which is armoured with cellulose plates, but accepts unarmoured larger and smaller algae. Copepods are sometimes able to grasp individual items with the mandibles. *Acartia tonsa* can use the grasping mechanism to prey upon nauplii of other copepods while simultaneously filtering algal cells.

9.3.2 Reproduction and dispersal

9.3.2.1 Asexual versus sexual reproduction

Asexual reproduction occurs among most phyla and may take place by two fundamentally different processes. First, embryos may develop parthenogenetically from unfertilized eggs and second, the body may divide as it grows. Parthenogenesis has evolved recurrently, and continues to do so in many independent lines, especially in those occupying terrestrial and freshwater habitats. Some parthenogenetic lines have lost the ability to reproduce sexually, and although benefiting in the short term from a higher potential rate of increase than sexual relatives (by omitting males that cannot give birth), these parthenogenetic animals lack the evolutionary potential associated with sex and so are much more prone to extinction than sexual lines. Other animals incorporate both parthenogenetic and sexual phases into their life cycles. Cyclical parthenogens usually reproduce sexually when growing conditions begin to deteriorate. The parents die and the progeny endure the ensuing harsh periods as dormant zygotes, or early embryos, protected within resistant capsules. When favourable growing conditions return, each embryo develops into a parthenogenetic female. This female reproduces parthenogenetically throughout the favourable growing period, forming a clone of genetically identical modules. Such a life cycle is similar to that of many unicellular algae (p. 227) and is associated with similar

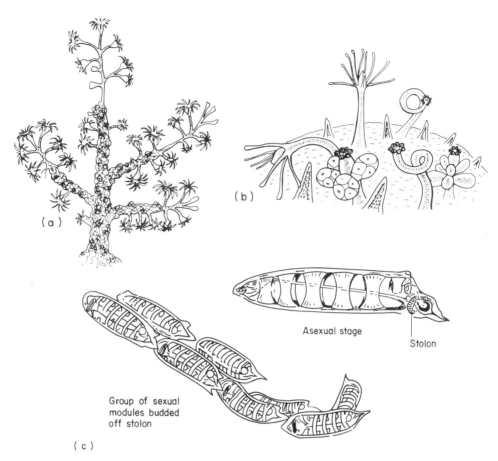

(a)

(b)

Asexual stage

Stolon

Group of sexual
modules budded
off stolon

(c)

Fig. 9.8 Clonal animals. (a) When modules (zooids) of a growing clone remain
united they form a colony, which is usually a benthic sedentary form as in this
example of the hydroid *Bougainvillea*. (b) The clonal modules are genetically
identical but may become specialized for different functions as in *Hydractinia
echinata* with feeding (long tentacles), defensive (cluster of knob-like tentacles)
and reproductive (large vesicles, ♂ left and ♀ right of colony) zooids. (c) Modules
of a growing clone may become detached as with most pelagic tunicates. Depicted
is the salp *Iasis (Salpa) zonaria*, which alternates between an asexual,
clone-forming stage and a sexual, outbreeding stage. The asexual stage buds off
chains of modules from a stolon. The modules remain attached in groups for a
while, during which time they are functional females. After giving birth to a
single asexual young, each female module changes sex and becomes detached
from the others, shedding sperm into the sea-water and perhaps fertilizing
another clone. (After Hardy 1962.)

advantages, i.e. the maintenance of a favourable surface area to
volume ratio by splitting the 'body' into modules, the ability to
disperse modules over a wide area in search of unpredictably located patches of resource, the rapid exploitation of local patches
of resource, and the lessening of the probability of extinction of
the genotype by spreading the risk of mortality among many independent modules. Cyclical parthenogens include aphids, rotifers and cladocerans, but whereas rotifers and cladocerans are
common in fresh water, they are relatively scarce in the sea,
being overshadowed by copepods. Why neither marine nor

236

freshwater copepods have developed any parthenogenetic lines remains a tantalizing, unanswered question.

Asexual reproduction by budding or fission of the body does not involve gametogenesis and is therefore quite different from parthenogenesis. Both processes, however, result in the formation of a clone of genetically identical modules. Budding or fission is common among many invertebrate phyla and the resulting clonal modules may remain attached to each other to form colonies, e.g. colonial hydroids, zooanthids, corals, bryozoans and colonial tunicates (Fig. 9.8a), or they may become detached to form a clone of aggregated or dispersed modules, e.g. solitary hydroids, scyphozoans, anemones, certain polychaetes and most pelagic tunicates (Fig. 9.8c). Division of the growing body into a clone of modules has several advantages. If the modules are dispersed the clone gains similar advantages to those of parthenogenetic animals. *Thalia democratica*, for example, is a warm-water, pelagic tunicate that completes the alternation of asexual budding and sexual reproduction within two days. It has about the shortest generation time of any metazoan and is thus well able to exploit local phytoplankton blooms (Heron 1972). If the modules form an aggregation or an organically united colony, the dispersal ability of the clone is severely reduced, but several other advantages remain that are of great significance among sedentary animals. First, filter-feeding devices such as the tentacular crowns of coelenterates, bryozoans and polychaetes work efficiently only below a certain size. Modular iteration preserves the optimal size of the feeding apparatus while allowing a continued increase in biomass. Second, modules can become specialized for different functions. Such 'division of labour' is seen most often in organically united colonies, e.g. the hydroid *Hydractinia echinata* in which different polyps are specialized for feeding, reproduction and defence (Fig. 9.8b), but may occur in modular aggregations, as with the anemone *Anthopleura elegantissima* in which peripheral polyps forgo sexual reproduction and use the energy to defend the reproductive polyps within. Division of labour and co-operation among modules of a clone are not altruistic, since only a single genotype is involved, but are analogous to the division of labour among the organs of a non-clonal animal. Third, a very flexible growth form is achieved, which can be modified to suit local conditions, e.g. topography of the substratum, direction of prevailing currents (Fig. 9.4c), presence of other organisms. Fourth, sedentary animals are at risk to predation by carnivorous grazers. Modules escaping predation can sustain damaged clone-mates until these are regenerated or replaced. Fifth, the ability to compete for and retain space on the substratum is increased by the collective effort of the modules.

Among the great variety of forms exhibited by sedentary modular colonies, Jackson (1979) recognizes six basic shapes: runners (linear or branching encrustations), sheets (two-dimensional encrustations), mounds (massive, three-dimensional encrustations), plates (foliose projections from the substratum), vines (linear or branching, semi-erect forms with restricted zones of attachment) and trees (erect, usually branching projections) (Fig. 9.9). Runners and vines are the most opportunistic forms, advancing quickly in a linear fashion to make temporary use of unoccupied space before being outcompeted by other growth forms that show increasing commitments to survival within their own areas of settlement (sheets < mounds < plates < trees). For example the bryozoan *Electra pilosa* adopts the runner growth form and opportunistically colonizes unoccupied patches on the surface of *Fucus spiralis*, but is eventually overgrown by *Alcyonidium hirsutum* and *Flustrellidra hispida*, which grow more slowly in any single direction but also advance two-dimensionally

Fig. 9.9 Growth forms of sedentary modular colonies: (a) runner—bryozoan *Electra pilosa*; (b) sheet—bryozoan *Alcyonidium polyoum* overgrowing vine—hydroid *Dynamena pumila* (after Stebbing 1973); (c) mound—sponge *Polymastia* sp. (after Bowerbank 1874); (d) vertical plate—sponge *Phakellia robusta*; (e) horizontal plate—sponge *P. ventilabrum* (after Bowerbank 1874); (f) tree—bryozoan *Bugula plumosa* (after Ryland 1962).

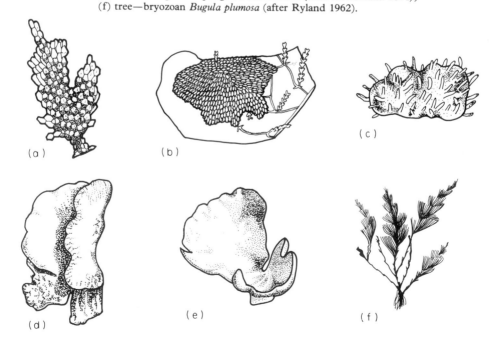

(a)   (b)   (c)

(d)   (e)   (f)

to form sheets over the substratum (p. 137). When two similar growth forms meet, the competitive outcome may depend on subtle factors such as whether two growing edges meet or whether one growing edge impinges on a non-growing edge.

Sometimes colonies react allelochemically to competitors. When different, and therefore genetically distinct, clones of the sponge *Hymeniacidon* grow into contact, they produce a substance that interferes with cellular adhesion. By means of this chemical warfare, one colony usually dominates the other, being reminiscent of the competitive hierarchies by mesenterial digestion found among corals (p. 173). Allelochemical interactions are common among sponges, bryozoans and tunicates inhabiting crevices, caves and shaded overhangs on coral reefs. Coexistence among these animals apparently is facilitated by the presence of competitive loops rather than hierarchies. For example, species A might overgrow species B which might overgrow species C, but species C is able to suppress species A by an allelochemical interaction. The occurrence of competitive loops in a community of potential competitors forms what Buss and Jackson (1979) call a competitive network, the complexity of which increases with the frequency of loop formation. Among the inhabitants of cryptic coral reef habitats, competitive loops are common and often permanent, whereas among the epiphytes of *Fucus serratus* they are infrequent and temporary (p. 137), and are probably absent among corals.

9.3.2.2   Larval life

Sexual reproduction in most marine animals results in the formation of larvae that persist for varying lengths of time before metamorphosing into the pre-adult stage. The range of larval sizes is narrow, invertebrate larvae being about 0.5–1.5 mm in overall diameter and fish larvae perhaps an order of magnitude larger. Various modes of larval life are possible. Larvae may be pelagic (planktonic) or non-pelagic (benthic), feeding (planktotrophic) or non-feeding (lecithotrophic), brooded or non-brooded, and intermediate or mixed modes also occur. It is difficult to generalize about the relative advantages of the different modes of larval life because many interacting selective forces may be involved, and because the experimental testing of ideas is hindered by the difficulty of measuring larval survivorship and dispersal in the field. At present it is possible only to list plausible hypotheses. Before doing so, it will be helpful to consider some general biological relationships.

(1) The larger the egg the longer is the embryonic develop-ment time from fertilization to hatching (Fig. 9.10). There may be two contributing factors to this correlation. First, larger eggs contain more yolk, which retards cleavage; some animals reduce this effect by providing extra-embryonic yolk supplies to the em-bryos, e.g. the non-developing 'nurse eggs' contained within the egg capsules of dogwhelks (*Thais lapillus*). Second, the longer the pre-hatching development time, the more advanced is the devel-opmental stage at hatching. Animals requiring developmentally more advanced hatchlings must therefore provision the eggs with more yolk, thereby making them larger. The largest eggs have sufficient yolk to nourish the embryo through to metamorphosis before hatching (direct development), whilst the smallest eggs contain yolk sufficient only for limited embryonic development, so that the hatching larvae must spend some time feeding in the plankton (planktotrophic) in order to complete their devel-opment. Eggs of intermediate size hatch into 'lecithotrophic' larvae, endowed with sufficient yolk to sustain them during their brief motile existence and to enable them to complete metamor-phosis. Postlarval juveniles hatching from larger eggs with direct development are slightly more advanced or larger than newly metamorphosed postlarvae from smaller eggs with lecithotrophic development. Among direct developing eggs, larger eggs produce larger hatchlings.

Fig. 9.10 The time taken by nudibranch eggs to develop and hatch as veligers from the egg capsules increases in larger eggs. (After Todd & Doyle 1981.)

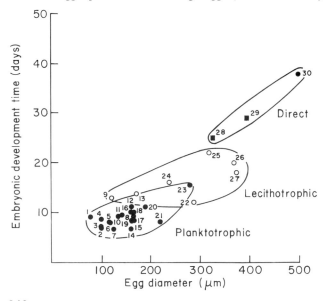

(2) Because the energy available to a parent for egg production is limited, the larger the egg the smaller is the clutch size (Fig. 9.11a). It follows from (1) and (2) that planktotrophic larvae are produced in greater quantities per brood than lecithotrophic larvae, which are in turn produced in greater quantities than direct developing larvae.

Fig. 9.11 (a) The fecundity of polychaetes within the *Capitella capitata* species-complex decreases as larger eggs are produced. (b) Egg size, and hence the amount of yolk reserve, is negatively correlated with the time that larvae spend in the plankton. *Capitella* species IIIa has direct development without a pelagic phase, species I, Ia and II have lecithotrophic larvae with short pelagic phases, and species III has plankotrophic larvae with a longer pelagic phase. (After Grassle & Grassle 1977.)

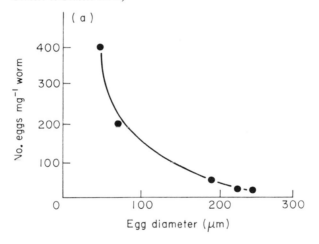

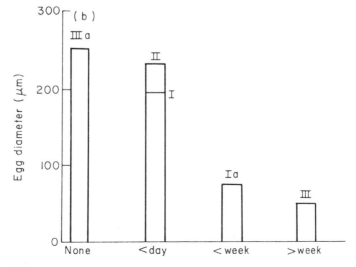

(3) Released larvae are at constant risk to predation, and if pelagic, also to transportation by currents into unfavourable areas. The cumulative risk of mortality therefore increases exponentially with increased duration of the mobile phase.

(4) Dispersal ability increases with increased duration of the mobile phase. This relationship is not simple, however, because in nearshore waters tidal currents are usually stronger than residual currents, with the result that the average distance transported would increase to a first maximum after six hours in the plankton, beyond which it would increase only very slowly. Hence a pelagic phase exceeding six hours duration would achieve little additional dispersal while greatly increasing the cumulative risk of mortality. Significantly increased dispersal would only be achieved by greatly prolonging the pelagic phase so that larvae could be transported long distances by non-tidal currents (see Crisp 1976).

With these generalizations in mind, some guesses can be made about the ecological significance of various modes of larval development.

Dispersal hypothesis

Animals with limited powers of postlarval dispersal, e.g. most benthic invertebrates, must rely on pelagic eggs and/or mobile larvae for dispersal. As with seeds transported by the wind, the small size of larvae enables them to use passive transportation (by water currents) as an energetically cheap means of dispersal. Also like seeds, larvae have some capacity to delay metamorphosis until a suitable settlement site is encountered. Among non-feeding larvae, energy reserves dictate that metamorphosis can be delayed only for a short while (Table 9.4), after which settlement becomes more indiscriminate and usually proves fatal. Feeding larvae can delay settlement for longer, but never as long as many seeds, which can remain dormant and viable for several years. Larvae on the other hand have the additional feature of small-scale locomotive powers, so that the encountering of suitable places for settlement and metamorphosis is less haphazard than with seeds.

Increased dispersal ability is achieved at the cost of increased larval mortality (see (3) above). Long-distance dispersal therefore has a reasonable chance of success only if very large numbers of larvae are released, and because of energetic constraints (see (2) above), these must be planktotrophic. As long as food is available to them, planktotrophic larvae theoretically could spend any

Table 9.4a Maximum time of survival in days at an assumed oxygen consumption of 5 ml g^{-1} (dry wt.) per hour. (After Crisp 1976.)

Source of energy	Percentage of tissue weight devoted to energy store				
	5%	10%	25%	50%	75%
Lipid	0.8	1.7	4.2	8.3	12.5
Protein	0.5	1.0	2.5	5.0	7.5
Carbohydrate	0.3	0.7	1.7	3.3	5.0
Metabolic rate of non-storage tissue ml O$_2$ g^{-1} per hour	5.3	5.6	6.7	10.0	15.0

Table based on $t = \dfrac{q \times x}{24 \times r}$ days.

q = oxygen requirement in ml g^{-1} of metabolite. X = fraction of body tissue devoted to reserves. r = respiration rate in ml g^{-1} per hour.

Table 9.4b Oceanic diffusion. (After Crisp 1978.)

Pelagic life	Log$_{10}$ (probable distance transported in cm)	Order of magnitude	Whether likely to be exceeded by tidal currents
3–6 h	4	100 m	Yes
1–2 days	5	1 km	Yes
7–14 days	6	10 km	?
14 days–3 months	7	100 km	No
1 year	8	1000 km	No

amount of time drifting, and indeed some Pacific echinoderm larvae spend as much as 36 weeks in the plankton, giving them sufficient time to traverse the ocean in favourable currents. If long-distance dispersal is unnecessary, an animal could gain by producing fewer larger eggs, each of which has a greater chance of surviving to metamorphosis, a trend which reaches its peak with direct development and elimination of the mobile larval stage. The optimal dispersal ability will depend on the spatial distribution and degree of permanence of suitable sites for colonization. If these are unpredictably located, then long-distance dispersal may be advantageous, whereas if they are predictably located nearby, then more limited dispersal will suffice and greater larval survivorship will be more advantageous. Also, predictably located settlement sites are more likely to be encountered by larvae than unpredictably located sites, so that competition among larvae and postlarvae may be more severe in the former than in the latter. Competitive ability will increase with advanced embryonic development or with larger size, so that increased egg size and associated reduced clutch size may be expected among animals successfully exploiting predictably located settlement sites.

Some support for the dispersal hypothesis is to be found in the reproductive modes of sibling species of polychaete comprising the *Capitella capitata* species-complex. These sibling species are morphologically so similar that their identities were only ascertained when their isozymes were examined electrophoretically by Grassle and Grassle (1977). Their morphological similarity and recent common ancestry makes these polychaetes ideal subjects for comparing variations in reproductive mode, because the influences of morphological and phylogenetic constraints can be discounted. Among the sibling species of *Capitella*, decreasing egg size is correlated with increasing clutch size, with increasing duration of the pelagic phase and hence with increasing powers of dispersal (Fig. 9.11b). Species IIIa has non-pelagic (benthic) larvae with very limited dispersal potential, species I and II have non-feeding (lecithotrophic), pelagic larvae, while the remainder have feeding (planktotrophic) larvae that remain in the plankton for several days (species Ia) to two weeks (species III) and are capable of long-distance dispersal. All the *Capitella* species can be regarded as opportunistic (p. 217), typically colonizing fine sediments that have been denuded of competitively superior species by environmental disturbance such as pollution. Some of the *Capitella* species, however, behave more opportunistically than others. Species I, II and IIIa are the most opportunistic, settling soon after hatching as relatively advanced larvae. Larval survivorship is high and these species can rapidly colonize local areas after disturbance has eliminated competitors. The continued existence of species I, II and IIIa depends on the frequent occurrence of disturbed patches of habitat, which they are ready to exploit at any time by virtue of their continued reproduction throughout the year. Species Ia and III occur in less variable subtidal habitats and because of the longer pelagic phase, they are slower to colonize new habitats than the more opportunistic species. The relatively wide larval dispersal of species Ia and III, however, enables them to select potentially more favourable habitats, which will be rarer and less predictably located than the frequent local disturbances exploited by species I, II and IIIa. Reproduction of species Ia and III is confined to a relatively narrow period in the winter and early spring. Perhaps this timing is related to optimal conditions for the planktotrophic larvae (food supply, survivorship) but other factors may be involved as discussed in the next hypothesis.

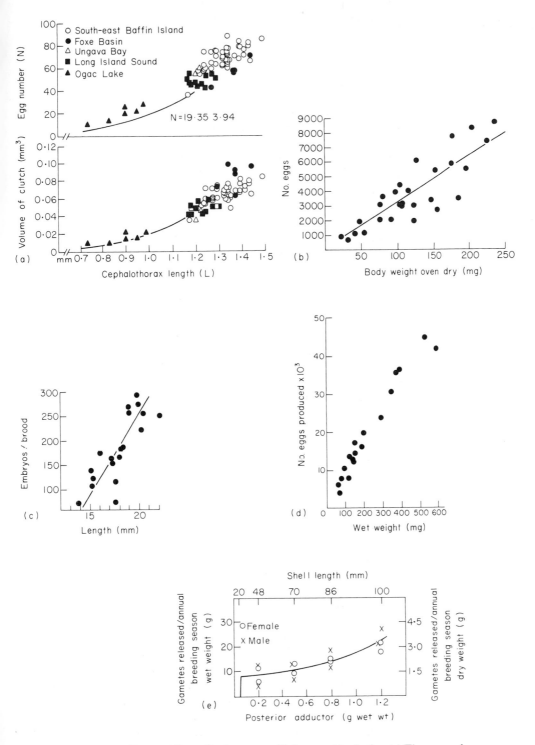

Fig. 9.12 Fecundity increases with increased body size. (a) The copepod *Pseudocalanus*. (After McLaren 1965.) (b) The barnacle *Tetraclita rufotincta*. (After Achituv & Barnes 1978.) (c) The isopod *Idotea baltica*. (After Strong & Daborn 1979.) (d) The polychaete *Harmothoe imbricata*. (After Daly 1972.) (e) The mussel *Choromytilus meridionalis*. (After Griffiths 1977.)

245

Settlement-time hypothesis

Some animals require larvae to settle not only in suitable locations but also at a certain time of the year. Such is the case when food for the postlarvae is present only for a brief seasonal period, which therefore sets the timing of settlement. Spawning time may be determined by several factors, but other things being equal, an animal should spawn when its reproductive value (expectation of future offspring) is at a maximum. Fecundity generally increases with body size (Fig. 9.12), but if an animal delays reproduction until it grows bigger it also runs the risk of being killed before spawning takes places. The age at maximum reproductive value therefore represents the optimal combination of age-specific fecundity and life-expectancy (Fig. 9.13), which will depend on growth conditions (temperature, food supply) and mortality factors (environmental, predation) prevailing from settlement to sexual maturation.

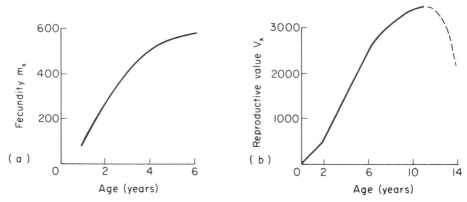

Fig. 9.13 (a) The fecundity of the periwinkle *Littorina neritoides* increases with age (i.e. size, see Fig. 9.12) and (b) the reproductive value (expectation of future daughters) peaks at an age when growth has almost ceased. (After Hughes & Roberts 1981.)

Onchidoris bilamellata is an annual nudibranch that feeds on intertidal barnacles. Postmetamorphic *O. bilamellata* are 0.5 mm in overall diameter and can handle only newly metamorphosed barnacle spat. Throughout most of Britain, *Balanus balanoides* is the principal prey and settles April–May. *O. bilamellata* larvae must also settle April–May so that the postlarvae can feed on *B. balanoides* spat before the latter grow too big to be handled. Growth and survival conditions on a North Yorkshire shore determine that the maximum reproductive value of local *O. bilamellata* occurs in late February, some 13 weeks before the required

settlement time. *O. bilamellata* can span this gap with planktotrophic larvae that metamorphose about 108 days after spawning.
If spawning remained at the optimal time, lecithotrophic or
direct developing larvae would metamorphose far too early and
the postlarvae would starve before barnacle settlement took place
(Fig. 9.14). Conversely, if lecithotrophic or direct developing
larvae were produced at a time when their metamorphosis would
coincide with barnacle settlement, the adult reproductive value
would be much reduced. This interpretation of the life history of
Onchidoris bilamellata, however, will remain tentative until it can
be shown that the timing of maximum reproductive value is
indeed fixed by environmental conditions and could not be delayed by modifications to the growth and behavioural patterns of
this species.

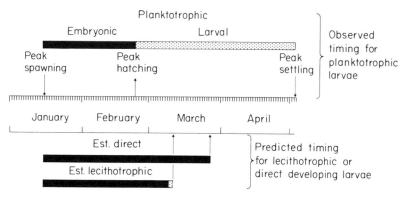

Fig. 9.14 The settlement-time hypothesis of Todd and Doyle (1981), proposed to
explain the timing of spawning in the nudibranch *Onchidoris bilamellata*.

In summary, the settlement-time hypothesis, proposed by
Todd and Doyle (1981), predicts that when settlement is restricted to a short period in the year by food availability or some
other critical factor, the mode of larval development may be determined by the time-gap between settlement and the most suitable time for spawning, other things being eqyal. Sometimes
other things are not equal, as discussed in the next hypothesis.

Size-threshold hypothesis

Some authors have postulated that in order to produce sufficient
numbers of lecithotrophic or direct developing larvae, an animal
must attain a threshold size, at which the body will contain sufficient energy reserves to manufacture the large eggs. Below the
threshold, the body contains insufficient stored energy to make

lecithotrophy or direct development worthwhile, because the resulting clutch sizes would be too small. Todd and Doyle (1981) use this hypothesis to explain lecithotrophy in the larger nudibranch *Adalaria proxima* and planktotrophy in the smaller nudibranch *Onchidoris muricata*.

Conversely, others have postulated that planktotrophy will not be feasible below a threshold body size because, even though the eggs are small, restricted bodily energy reserves could not produce enough of them to compensate for the larval mortality incurred by a long pelagic phase. Underwood (1978) has combined both ideas by postulating an upper threshold body size above which lecithotrophy is more advantageous, and a lower threshold below which lecithotrophy again is more advantageous, planktotrophy being the optimal development mode between the two thresholds. Theories that seem to demand the best of all possible worlds do not instil much confidence, but there is circumstantial evidence for a threshold body size below which direct development is more advantageous than the production of pelagic larvae. Tiny bivalves belonging to such genera as *Gemma* and *Mysella*, which live in shallow-water sediments, or *Lasaea* and *Turtonia*, which inhabit small intertidal crevices, all brood their embryos and release them at the postlarval stage. *Lasaea rubra*, for example, incubates 12–22 embryos in its suprabranchial chambers and releases them as 0.5–0.6 mm juveniles, even though the parent is itself only 2–3 mm in length. Perhaps, with such small energy reserves, the potential clutch size of these tiny animals is so restricted that larval mortality must be avoided altogether. This also may be the case among meiofaunal animals, ranging from 50 μm–3 mm in length and adapted to life in the interstices of sedimentary particles. Ninety-eight per cent of interstitial species lack pelagic larvae and many brood their young to an advanced developmental stage, while others protect their eggs in sticky cocoons that readily adhere to sand grains. Clutch sizes are correspondly small, two to three eggs being normal and seldom exceeding ten.

Energy-subsidy hypothesis

The plankton is a food resource used by animals ranging from small invertebrates to the largest vertebrate (Chapter 2). During phytoplankton blooms this food resource is particularly rich and it is conceivable that it may pay some species to exploit the plankton not only as food for themselves but also as food for their progeny, thereby subsidizing the energetic cost of reproduction.

Fecundity would be increased because the parent need make only a small energetic investment per larva (Fig. 9.11b). Larval growth, and hence chance of successful recruitment, would be enhanced when planktonic food is particularly abundant, allowing the parent to increase its reproductive value in years with a good bloom. The converse, however, would be true in years with a poor bloom, and it has not yet been demonstrated that there is a net gain in energy as a result of feeding in the plankton. It is certain that the spawning of some animals is timed so that their larvae can exploit the phytoplankton bloom. For example, increased phytoplankton concentration in the spring triggers spawning in the chiton *Tonicella lineata* on western North American shores.

Food-niche hypothesis

It has been suggested that by exploiting a different food resource, planktotrophic larvae avoid competition with the parental generation, but such competition could equally be avoided by the production of non-feeding larvae. Feeding larvae may need to forage in the plankton to encounter suitably small food organisms, but again this constraint could be avoided by provisioning the young with yolk. The food-niche hypothesis therefore cannot explain the relative selective advantages of feeding and non-feeding larvae among different animals.

Although larvae are chiefly pelagic in low latitudes where planktonic productivity is less seasonal, direct development is prevalent in polar regions where planktonic production is highly seasonal and irregular, making planktotrophy too risky (see p. 82).

9.3.2.3 Parental investment

Planktotrophic, lecithotrophic and direct developing larvae represent increasing levels of energetic investment per offspring by the parent. Some of the possible selective advantages of these different levels of parental investment are discussed in the previous section, but the picture is left far from complete. Among species with direct development, parental investment varies considerably according to the amount of yolk provided per egg, the amount of encapsulating material per egg and the amount of parental care in the form of brooding or guarding the eggs and young (Fig. 9.15). Parental investment beyond the minimum required to produce a viable egg is worthwhile only if juvenile survivorship is significantly increased and outweighs the concomitant

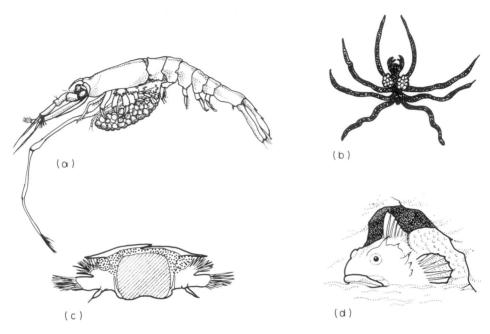

Fig. 9.15 Parental care. (a) As is typical among decapods, the euphausiid, *Nematoscelis difficilis*, carries its eggs on modified thoracic limbs. (After Briggs 1974.) (b) In pycnogonids (Arachnida), females transfer their eggs to males, which carry the eggs on modified limbs. (After Nakamara & Sekiguchi 1980.) (c) The scale worm, *Harmothoe imbricata*, broods its eggs on its back, where they are covered by plate-like processes from the parapodia. (After Daly 1972.) (d) The sea scorpion (sculpin), *Cottus bubalis*, guards its egg mass. (After Hughes 1980b.)

losses in potential fecundity or dispersal potential. For example, the winkle *Littorina rudis* retains its young within a brood chamber, which is a modified jelly gland, until the shell is well developed and the young are able to lead an independent existence on the shore. *Littorina nigrolineata* has a functional jelly gland which it uses to provide a gelatinous protective capsule round the egg mass placed beneath stones or within crevices on the midshore. Because *L. nigrolineata* does not brood its eggs, these can be produced more quickly, so that the fecundity of *L. nigrolineata* is about three times that of *L. rudis* on the same shore. *L. nigrolineata* therefore thrives well at midshore levels where the survivorship of egg masses is high, but at higher shore levels egg masses would become desiccated and it is here that the brooding behaviour of *L. rudis* places the latter species at an advantage. Greater parental investment therefore enables *L. rudis* to reproduce in harsher environments (high-shore levels, salt-marshes) than related species which do not brood their young.

9.3.2.4   Semelparity and interoparity

Organisms may reproduce sexually once and then die (semelparity) or reproduce more than once, or continually, over a pro-
tracted period (iteroparity). Care should be taken not to confuse
semelparity and iteroparity, which are defined relative to the lifetime of the organism, with the terms ephemeral, annual and perennial, which are defined relative to the year. Nudibranchs, such
as *Onchidoris* spp. (p. 246) are annuals that die after spawning
and are therefore semelparous. Eels are perennials that also die
after spawning and are semelparous. Winkles, *Littorina* spp.
(p. 250), are perennials that are iteroparous, while some bryozoans, such as *Celleporella hyalina*, can be ephemeral and iteroparous.

Whether semelparity or iteroparity is the more advantageous
depends very much on the ratio of juvenile to adult survivorship,
as determined by the morphologies and habitats of the two
stages. Semelparity relies on the survival of the young to maturity whereas iteroparity relies more on survival of the adult. The
benthic adults of marine, sedentary invertebrates are less at risk to
mortality than the dispersing larvae and iteroparity predominates
among these organisms. Apart from some very small species,
most prosobranch gastropods are well protected by their shells
and have a much higher life expectancy as adults than as juveniles, and are correspondingly iteroparous. Nudibranchs lack a
shell and even though they use other protective devices such as
acidic secretions, nematocyst-bearing dorsal cerata or crypsis,
they are likely to have lower life expectancies during the benthic
stages than have the armour-plated prosobranchs. All nudibranchs are semelparous. These rationalizations should be regarded with caution, however, because natural magnitudes of
juvenile/adult survivorship necessary for a sound comparison
have not been measured.

Semelparity in migratory fish such as salmon, eels and lampreys probably has a different evolutionary history. The life
history of salmon and lampreys is partitioned between marine
habitats that are highly productive, promoting growth and hence
fecundity of the adults, and freshwater habitats that are much
less productive, but safer nursery grounds for the young. The
migration from feeding to nursery grounds, however, demands so
much energy and entails such high risks that it pays only to do it
once, committing all the remaining energy into a suicidal bout of
reproduction. The reverse migration of eels, from freshwater
feeding grounds to marine spawning grounds, is more difficult to

understand. Eels often exploit productive, relatively safe habitats such as eutrophic muddy ponds, but the conditions on the nursery grounds in the Sargasso Sea remain unknown. Perhaps it is significant that eels have a recent marine ancestry and salmon a freshwater one.

9.3.2.5   Hermaphroditism, why combine sexes?

In most actively mobile animals male and female functions are performed by separate individuals, termed gonochorists (equivalent to dioecious plants), and in a minority by single individuals, termed hermaphrodites (equivalent to monoecious plants). In simultaneous hermaphrodites male and female functions are performed together, whereas in sequential hermaphrodites the male function may precede the female function (protandry) or vice versa (protogyny). How can these sexual permutations and combinations be accounted for in terms of natural selection?

Gonochorism owes its origins to the evolution of anisogamy. Primitive sexual organisms probably had one type of gamete (isogamy) as does the unicellular alga *Chlamydomonas*. Fertilization and zygotic development require mobility, whereby unrelated gametes can encounter one another, but also energy reserves to fuel subsequent development. Mobility is facilitated by small size but hindered by bulky yolk, hence the evolutionary division of labour (anisogamy) into smaller motile sperm whose function is to seek, and larger non-motile eggs whose function is to nourish. It will often be advantageous for division of labour to be carried through to the parents, males being specialized for seeking females and perhaps defending them from other males (ensuring paternity), females being specialized for nurture of the young. Different adult morphological features are often involved, especially with regard to the reproductive organs. The combination of sexual roles within a single individual could therefore be disadvantageous because of incompatible morphology and because of the extra energetic investment in two sets of reproductive apparatus. Hermaphrodites are, however, widespread in nature, so that in certain circumstances the advantages of gonochorism must be outweighed by other factors.

A clue to one selective advantage of hermaphroditism is given by the greater preponderance of sedentary than of mobile hermaphrodites. Among gonochoristic, sedentary animals, there is a risk that neighbours may be of similar sex and unable to fertilize each other, a risk that is avoided by hermaphroditism. Similarly, mobile animals that normally occur at low densities are sometimes hermaphroditic, e.g. nudibranchs, thereby ensuring that when two mature individuals meet, they are capable of cross-fertilization. An interesting alternative solution to this problem is found among several gonochoristic, oceanic angler-fish, e.g. *Ceratias holbolli*, which live at exceedingly low population densities: the dwarf male becomes organically attached to the female, deriving nourishment from her and remaining with her throughout reproductive life. The rate of encounter between sexes is so low that once contact has been made, it is advantageous to maintain it.

Whereas hermaphroditic animals that copulate, such as barnacles and nudibranchs, function simultaneously as males and females (Fig. 9.16e), sedentary animals that liberate sperm into the sea-water are usually sequential hermaphrodites, e.g. the

Fig. 9.16 Hermaphrodites. (a) A protogynous sequential hermaphrodite, the colonial tunicate, *Botryllus schlosseri*. (After Millar 1970.) (b) A protogynous sequential hermaphrodite, the slipper limpet, *Crepidula*. (c) A protandrous sequential hermaphrodite, the limpet, *Patella vulgata*. (d) A protandrous sequential hermaphrodite, the shrimp, *Pandalus montagui*. (After Smaldon 1979.) (e) A simultaneous hermaphrodite, the nudibranch, *Archidoris pseudoargus*. (f) A simultaneous hermaphrodite, the barnacle, *Balanus balanoides* (showing copulation).

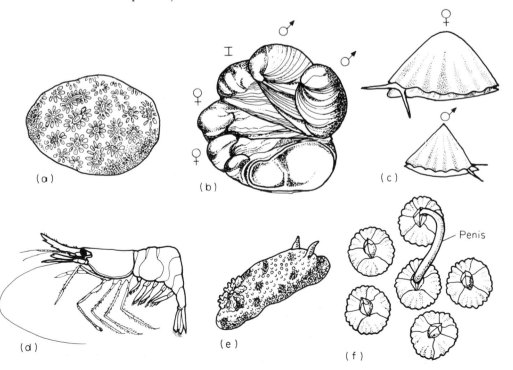

protogynous tunicate *Botryllus schlosseri* (Fig. 9.16a), thus preventing self-fertilization and loss of fitness due to homozygosity.
The cyclical, sequential hermaphroditism of sedentary animals,
however, is rather different from the permanent, sequential hermaphroditism of certain mobile animals. In populations of the
limpet *Patella vulgata* most individuals become male at a relatively small size and switch to being female as they grow larger
(Fig. 9.16c). A size threshold has been postulated, rather like that
for planktotrophy (p. 247), below which energy reserves are in-
adequate to produce a sufficient number of eggs (that hatch into
planktotrophic larvae). Sperm, however, are energetically cheap
to produce, so that small individuals can function adequately as
males. Difficulties arise with this interpretation because of inter-
male competition. Bigger males would be more fecund than
smaller males, and since limpets are external fertilizers, the larger
males should outcompete smaller ones to fertilize eggs. Indeed, a
small proportion of *P. vulgata* do remain male throughout life.

Protogyny is common among coral-reef fish. In some wrasses
and parrot-fish, only territory-holding males are accepted as
mates by the females. Male reproductive success therefore depends on being able to defend a territory from competing males
and this can only be achieved by large, strong fish. Young, small
fish could not compete for territories, so they function as females
until the critical size is reached.

Sometimes, sex is determined environmentally. The coral-reef
fish *Anthias squammipinnis* lives in schools within territories on
the reef. Each school is comprised of females attended by a male.
If the male is removed, one of the females changes sex, evidently
in response to visual behavioural stimuli. The value of this particular sex ratio, however, remains to be elucidated. Some species
of shrimp, belonging to the genus *Pandalus*, are protandrous
(Fig. 9.16d), but can adjust the age of sex change according to
the age composition of the population which fluctuates from year
to year because of irregularities in recruitment. This flexibility is
selectively advantageous because the value to an individual of
being male rather than female increases when males are rarer. As
first pointed out by Fisher (1930), on average the rarer sex con-
tributes genetically to more zygotes than the commoner sex.

9.3.2.6   Reproductive effort

Different modes of larval development and parental care represent various ways in which an animal can spend the resources
(energy, nutrients, time) allotted to sexual reproduction, but

what factors determine the total amount of reproductive expenditure? For simplicity and comparability energy can be regarded as the resource of overriding importance. Energy assimilated from the food is either accumulated as body tissues and gametes, dissipated as heat resulting from metabolism, or lost in nitrogenous excretions, leakage of dissolved organic matter, or in the production of mucus. Usually over 50% of assimilated energy is lost as heat, but the amount varies according to the level of metabolic activity, and may exceed 80%. Muscular activity and physiological homeostatic mechanisms such as osmoregulation are metabolically costly. Since in the long term an animal assimilates energy at a fixed rate, increased metabolic costs must reduce somatic growth or gametogenesis, and vice versa. The proportion of assimilated energy devoted to reproduction is termed *reproductive effort*, and this may be expected to vary according to the life history and the degree of environmental stress. In the strict sense, reproductive effort includes not only the energy accumulated in gametes and associated structures such as egg capsules, but also the metabolic energy spent on all reproductive activities such as searching for mates, copulation, defence of young and other kinds of parental care. In practice, reproductive metabolic costs are extremely difficult to measure, so that estimates of reproductive effort are usually approximations in the form of the proportion of assimilated energy channelled into eggs or even just the ratio of egg production to somatic biomass. The pattern emerging from such measurements is complex, but there is a marked tendency for total lifetime reproductive effort to be greater in semelparous than in iteroparous animals, and for 'instantaneous' reproductive effort to increase with increased age and decreased life expectancy in iteroparous species (Fig. 9.17).

Fig. 9.17 Reproductive effort (the proportion of assimilated energy used in spawn production) is greater in semelparous (s) than in iteroparous (i) marine snails. (From Hughes & Roberts 1980.)

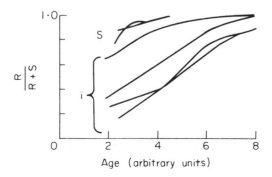

Age (arbitrary units)

Explanations for these trends are similar to those for semelparity and iteroparity (section 9.3.2.4). If the body is unlikely to survive beyond a single reproductive season, then once sexual maturity is reached, all available energy should be committed to reproduction, depleting the body reserves to a lethal level. If the body is likely to survive beyond a single reproductive season, it becomes worth investing a greater proportion of assimilated energy in bodily maintenance. The likelihood of such an investment paying off, however, declines as life expectancy decreases, so that less energy should be gambled on the body and more should be firmly committed to reproduction as the animal ages. More opportunistic (r-selected) species may be expected to have higher reproductive efforts than more stable-habitat (K-selected) species (see section 9.1 and Pianka (1974)).

10 Speciation and Biogeography

10.1 Introduction

Within an ecosystem, such as a coral reef, rocky shore, or tract of sediment, the numbers and kinds of organisms present are determined by the availability of resources, competition, predation, environmental disturbances, local colonizations and extinctions, all acting on an ecological time scale and therefore having an immediate influence on the biota (Chapters 1, 3, 5, 6, 7). The possible combinations of species present, however, will depend on the pool of potentially available species in the geographical area. Guilds of herbivorous grazers on rocky shores are dominated by periwinkles, top shells and limpets in temperate regions, but by a wider range of gastropods together with grapsid crabs on tropical shores (p. 117). Shallow, sublittoral, rocky substrata are colonized by kelps in temperate latitudes and by corals in the tropics (Figs. 5.12, 6.3). Geographical faunistic and floristic variations are determined not only by immediate ecological and environmental factors, but also by processes of speciation, colonization, and extinction acting on an evolutionary time scale. Of course, ecological and evolutionary time scales merge, and at a detailed level are meaningful only with reference to the generation times of particular organisms. Fruitful generalizations can be made, however, and the following examination of speciation and biogeographical patterns forms a natural extension of the ecological discussions in previous chapters.

10.2 Speciation

10.2.1 Allopatric speciation

The number of species living on Earth represents a balance between rates of speciation and extinction. Causes of extinction are little understood, but somehow must be related to changes in the physical environment, to changes in the biological environment as competitors, predators and pathogens evolve, or, in small isolated populations, to accidents of chance. From the fossil record, it appears that extinctions tend to occur in waves rather than gradually, suggesting the influence of climatic changes.

Speciation occurs when an ancestral gene pool diverges along two or more lines in response to different selective regimes and gene flow eventually ceases between the divergent lines. Divergence is facilitated by the spatial isolation of gene pools. Different areas are likely to produce different selective regimes and spatial separation will tend to reduce genetic mixing of the populations. If by subsequent migration the populations meet, interbreeding may produce hybrids with incompatible genetic combinations, resulting in low fitness. In this case, reproductive isolating mechanisms will be selected for and separate species created. The formation of species from a geographically fragmented ancestral gene pool is called *allopatric speciation*.

Incipient allopatric speciation is occurring among populations of the small harpacticoid copepod *Tisbe clodiensis* from different parts of the Mediterranean and Atlantic coasts of Europe. Individuals from most populations interbreed in the laboratory, but this results in progeny of reduced viability. Evidently the geographically isolated gene pools have become modified by local conditions so that when alien gametes combine, the finely tuned genetic architectures break down, resulting in hybrids of low fitness. Reproductive isolating mechanisms have not evolved between these disjunct populations, and indeed would only be expected to occur if the populations had become sympatric.

Such an event has evidently happened among German populations of small littoral isopods belonging to the *Jaera albifrons* species complex. By allowing individuals access only to mates from other species of the *albifrons* complex, hybridization can be induced in the laboratory and although the F_1 progeny are viable, the F_2 progeny show reduced viability. In nature, the frequency of hybrids is less than 1%, due to the operation of behavioural, reproductive isolating mechanisms between sympatric populations. Mating is preceded by courtship, and females refuse males presenting alien stimuli.

If, as is thought to be the case, allopatric divergence of gene pools is the principal mechanism of speciation, then the rate of speciation ought to be highest in the most heterogeneous environments, where opportunities for spatial isolation and divergent selection are greatest. These opportunities are less likely to arise in the water column than on the sea bed, and accordingly only about 2% of marine species are entirely pelagic. Environmental conditions within the interstices of sediments are more monotonous throughout the world than those on exposed surfaces. Consequently, geographical variations are far more pronounced among epifaunas than among infaunas. Gastropods, for example, tend to have restricted geographical distributions and entirely different species live in the tropics than in temperate regions. Infaunal polychaetes on the other hand, tend to have wide geographical limits and many are cosmopolitan.

Not only the physical nature of the environment, but also the dispersability of organisms will influence the opportunity for

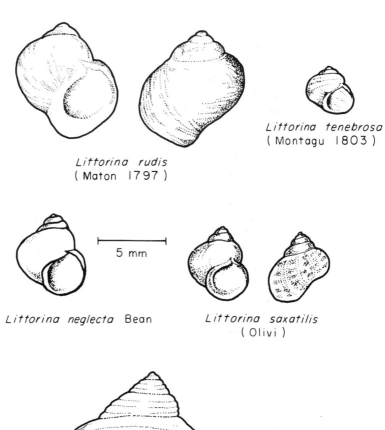

Littorina rudis
(Maton 1797)

Littorina tenebrosa
(Montagu 1803)

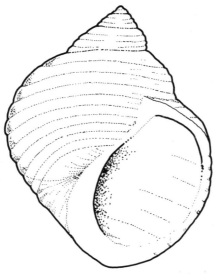

5 mm

Littorina neglecta Bean

Littorina saxatilis
(Olivi)

Littorina nigrolineata Gray

Fig. 10.1 *Littorina saxatilis* species-complex. *L. rudis*, *L. nigrolineata* and *L. neglecta* are accepted as distinct sibling species. *L. tenebrosa* and *L. saxatilis* may be morphs of *L. rudis*. (After Fretter & Graham 1980.)

speciation. Gene flow is more restricted among animals with direct development than among those with pelagic larvae, and more restricted among those with short-lived than among those with long-lived, pelagic larvae. For example, periwinkles in the *Littorina saxatilis* species-complex brood their young or lay benthic egg masses from which crawling young emerge. Dispersal is therefore almost entirely by crawling and perhaps occasionally by rafting on floating objects, so that gene flow between distant populations must be very slight indeed. On European shores, at least three species and several other morphs of undetermined taxonomic status exist within the *saxatilis* complex (Fig. 10.1). As expected, electrophoresis of allozymes shows that there is considerable genetic variation among populations in the *saxatilis* complex but that populations of *Littorina littorea*, which has planktotrophic larvae, are genetically more homogeneous even from both sides of the Atlantic ocean.

The production of pelagic larvae, however, does not necessarily lead to thorough genetic mixing. Inside Long Island Sound is a population of *Mytilus edulis* genetically distinct from populations outside, as revealed by the electrophoresis of allozymes (Lassen & Turano 1978). The Sound has a typical estuarine hydrography, whereby less saline water flows out on top of a more saline, compensating bottom current. Larvae of resident mussels are carried out of the Sound in the surface current, but oceanic larvae fail to penetrate the Sound because when freshwater discharge is low during the summer breeding season, the weakened bottom current fails to flow over a sill near the mouth of the Sound. Genetic isolation of the Sound population is therefore imposed by hydrographical conditions.

A consequence of reduced dispersability and hence reduced interpopulation gene flow is that gene pools can become closely adjusted through natural selection to local conditions, but this is also likely to make them more vulnerable to environmental changes and prone to extinction. Enhanced dispersability and extensive interpopulation gene flow prevent localized differentiation of gene pools, so that species with long-lived, widely dispersing larvae must be broadly adapted to a range of environmental conditions and therefore less liable to extinction. It is sometimes possible to deduce the larval mode of life from the larval shell of fossilized gastropods. Scheltema (1978) thus deduced that the mean evolutionary longevity of species with teleplanic larvae (i.e. living several months to a year in the plankton) is about 19 million years, whereas that of species with direct development is only about three million years.

10.2.2 Sympatric speciation

At least in theory, geographical isolation is not always a prere-
quisite for speciation. When different selective forces act on
different parts of a continuous population, they may, if strong
enough to override the effect of gene flow, cause divergence
within the gene pool, eventually leading to reproductive isolation. Perhaps the easiest kind of situation to envisage is where
an ancestral gene pool diverges in response to the extreme conditions at opposite ends of a strong environmental gradient.

A difficulty with the concept of sympatric speciation is that
very high selection pressures would be needed to counteract the
homogenizing effect of gene flow. Attempts to cause sympatric,
genetic divergence in laboratory cultures of *Drosophila* have met
with varied success and only rarely has convincing evidence of
sympatric, genetic divergence and reproductive isolation been
found in the field (e.g. copper-tolerant plants on copper-mine
tailings). Genetic responses to strong selection pressures associated with environmental gradients are frequently found in
nature, but gene flow usually persists, preventing speciation and
resulting in gradual changes in gene frequency (clines) along the
gradients. For example Schopf and Gooch (1971) found that the
frequencies of two alleles at a leucine-amino-peptidase (Lap)
locus in the bryozoan *Schizoporella unicornis* changed along the
coastline of Cape Cod, parallelling a gradient in summer water
temperature (Fig. 10.2).

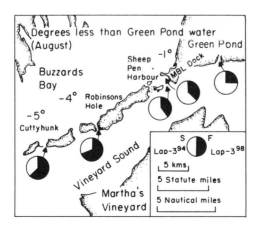

Fig. 10.2 Cline in frequencies of 'slow' and 'fast' alleles at a leucine–amino–
peptidase locus of *Mytilus edulis* associated with a temperature gradient. 'Slow'
and 'fast' refer to the relative speeds at which the proteins marking different
alleles move during electrophoresis. (After Schopf & Gooch 1971.)

Williams (1977) interpreted the data differently, proposing that there was probably sufficient mixing of larvae by currents to prevent any local genetic responses within the coastal population of *S. unicornis*. Larvae with a similar range of genetic qualities would be produced all along the coast each year. The clinal changes in gene frequencies could then only be caused by recurrent environmental elimination of locally inappropriate genotypes.

Without knowing the details of evolutionary history it is difficult to judge whether closely related, coexisting species have evolved sympatrically or have evolved allopatrically and subsequently become sympatric by range extension. However, the rapidly increasing number of sympatric 'sibling' or 'cryptic' species that are being discovered with the aid of electrophoresis of allozymes and other detailed taxonomic methods suggest that allopatric speciation cannot account for all cases. Sibling or cryptic species are groups of species that are morphologically very similar or even indistinguishable within each group (see p. 244), yet do not interbreed. Examples are the European sibling species *Littorina obtusata* and *L. mariae* that live on midshore fucoids and low-shore fucoids or kelps respectively, and cryptic species of the *Alcyonidium gelatinosum* complex (Fig. 10.3) discovered in dredge samples from a rocky bottom in the Bristol Channel.

10.2.3 Environmental harshness and instability

Whether allopatric or sympatric processes are involved in speciation, genetic diversity will be enhanced by environmental heterogeneity. But other factors can also be important. Harsh environments (e.g. extremes of temperature or salinity) and

Table 10.1 Mean genetic variation in species in three different environments. (After Battaglia *et al.* 1978.)

Environment and species / Parameter	Marine Tisbe clodiensis Tisbe holothuriae* Tisbe biminiensis	Brackish water Tisbe holothuriae** Gammarus insensibilis	Rock pools Tigriopus brevicornis Tigriopus fulvus
Mean number of alleles per locus	1.982	1.788	1.297
Mean percentage of loci polymorphic (0.99)	56.84	42.03	18.27
Mean percentage of loci polymorphic (0.95)	46.78	38.89	16.13

* Banyuls-sur-Mer.
** Sigean.

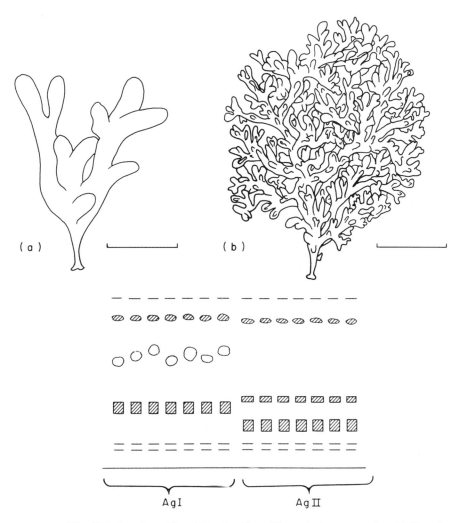

Fig. 10.3 Cryptic species within the *Alcyonidium gelatinosum* complex. (a) Type I; (b) type II; (c) esterase zymograms for types I and II. Scale bar = 3 cm. (After Thorpe *et al.* 1978)

fluctuating environments will cause strong selection for physiological robustness or phenotypic plasticity rather than genetic variability. Genetic variability within several species of crustacean on European shores decreases from marine, brackish-water to rock-pool habitats as the environment becomes harsher and more variable (Table 10.1).

Environmental variability hinders the fine tuning of genotypes to local conditions and is of particular significance with regard to fluctuations in food supply. Stable food resources allow competition to generate selection for trophic specialization. Fluctuating food resources render trophic specialization less feasible because animals will need to become more opportunistic in times of food shortage. Among species of krill, levels of genetic variation are correlated with the trophic stability of the environment. Heterozygosity increases in the order *Euphausia superba* from highly seasonal Antarctic waters, *E. mucronata* from Chilean waters with moderately irregular upwellings, and *E. distinguenda* from the trophically stable waters of the equatorial eastern Pacific. *E. superba* is apparently represented by relatively homogeneous generalist genotypes and *E. distinguenda* by more heterogeneous specialist genotypes.

10.3 Biogeography

10.3.1 *Latitudinal gradients of species diversity*

The latitudinal decrease in species diversity from equatorial to polar regions was considered briefly in Chapter 3, but can now be considered in further detail in the context of the determinants of speciation.

10.3.1.1 Area effect and environmental heterogeneity

The diversity of habitats and opportunities for the genetic isolation of populations on continental shelves increases as the total area of shelf increases. Therefore, the amount of speciation among shelf faunas ought to be higher in parts of the world with greater areas of shelf, assuming, as seems likely, that there is no compensating increase in extinction rate. The species diversity of bivalves and bryozoans is indeed correlated with continental shelf area, which decreases away from the equator, but there is much residual variation (Fig. 10.4). Schopf *et al.* (1978) suggest that the correlation between species diversity and shelf area would improve if total shelf area could be replaced by a measure of habitable shelf area. The proportion of shelf suitable for bivalves and bryozoans may vary independently of latitude. For example, the south-eastern coast of North America is sandy and more suitable for infaunal bivalves than the south-western coast, which has coarser sediments and more rocky areas favouring epibenthic animals such as bryozoans.

10.3.1.2 Trophic stability

Prosobranch gastropods are one of the most diverse groups of marine macroinvertebrates and they show a strong latitudinal

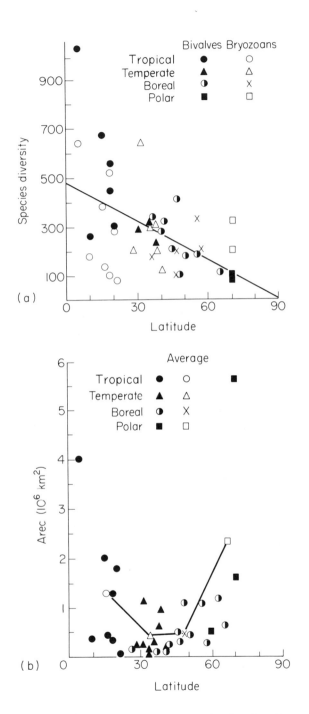

Fig. 10.4 (a) Latitudinal changes in species diversity of bryozoans and bivalves. (b) Latitudinal changes in a continental shelf area. (After Schopf *et al.* 1978.)

decrease in diversity from equatorial to polar regions. This trend may partly be due to latitudinal changes in the area of habitat, as with bivalves and bryozoans, but is also due to the trophic stability of the environment (Taylor & Taylor 1977). A sharp drop in the diversity of eastern Atlantic gastropods occurs round about latitude 40°N, coinciding with a change from a tropical oceanic regime in which primary production is continuous throughout the year, to a temperate oceanic regime in which primary production is seasonally pulsed. The trophic stability of tropical habitats allows specialized feeding behaviours to evolve, whereas at higher latitudes trophic instability has the opposite effect. Tropical families of predatory gastropods have narrower diets than families from higher latitudes. For example the Cassidae feed on echinoids, the Tonnidae on holothurians, the Mitridae on sipunculan worms, most Conidae, Vasidae and Bursidae on polychaetes, and the Harpidae on decapod crustaceans. By contrast, the Buccinidae, which are abundant at high latitudes, are opportunistic feeders, whose diets include polychaetes, sipunculans, crustaceans, molluscs, echinoids and carrion. A given supply of food resource can either be portioned among many species with narrower diets and smaller populations or among fewer species with wider diets and large populations, so that more species exist in low latitudes where diets are relatively narrower than at higher latitudes where diets are wider.

The Turridae are an exception which proves the rule of trophic stability. These predators are abundant at high latitudes yet appear to be specialist feeders on polychaetes. Their prey are, however, deposit-feeding worms whose population dynamics are little affected by the seasonally pulsed primary production.

Quality of the food resource changes throughout the world and this influences faunal species composition rather than species diversity. The proportion of non-predatory gastropods rises between latitudes 40–60°N (Fig. 10.5). This is due to a surge in the number of algal-grazing species in response to the high benthic algal productivity of temperate regions (Chapter 5).

10.3.1.3 Diversity associated with coral reefs

Due to the physiology of the coelenterate–dinoflagellate symbiosis, reef-building corals are confined to shallow, clear waters within the 20°C winter isotherms (Chapter 6). The intricate three dimensional architecture of corals and associated reef structures creates multitudinous microhabitats that support very rich faunas, an effect which is reinforced by the extremely high, stable

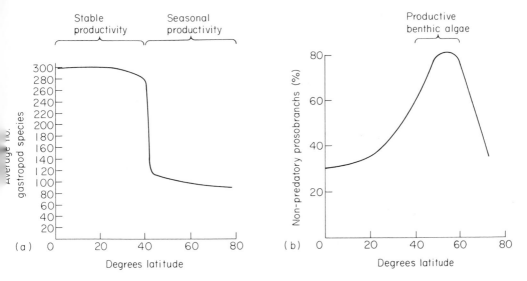

Fig. 10.5 (a) Latitudinal changes in the species diversity of gastropods in the eastern North Atlantic. (b) Latitudinal changes in the percentage of gastropods represented by algal grazers. (After Taylor & Taylor 1977.)

productivities of coral-reef ecosystems. Coral reefs, therefore, account for a significant proportion of the high overall species diversity of tropical shelf faunas.

10.3.2 Oceanic differences in species diversity

The highest diversity of shallow benthic species in the world is in the tropical Indo–Pacific, followed by the tropical Pacific coast of America. The lowest tropical benthic species diversities occur on both sides of the Atlantic. Within the Indo–Pacific the highest diversity is centred on the Indo–Malayan region (Fig. 10.6), where most of the diversity-promoting factors previously described occur together. First, the equatorial location is associated with stable, non-seasonal, primary production (section 10.3.1.2). This stability is enhanced by the numerous small islands and continental land masses surrounded by a large ocean, a geographical configuration that produces a very even maritime climate and stable water column. Second, the continental shelf is dissected by numerous deep basins between the islands and continents, providing good opportunities for the genetic isolation of populations (section 10.3.1.1). Third, the coral reefs provide a spatially heterogeneous environment and stable high primary productivity (section 10.3.1.3).

267

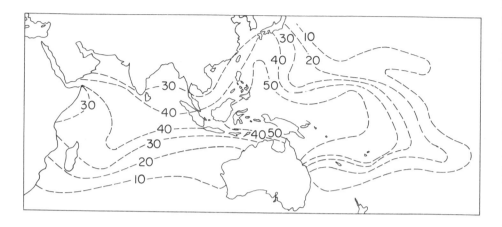

Fig. 10.6 Generic diversity of hermatypic corals in the Indo-Pacific. The centre of diversity is in the Indo-Malayan region. (After Stehli & Wells 1971.)

The reduced species diversity of shallow, benthic assemblages outside the Indo–Pacific region is partly correlated with the increased continentality of the climate. As a result of land having a lower specific heat than water, temperature fluctuations increase as the ratio of continuous land surface to sea surface increases. This ratio is least among the archipelagos and peninsulas of the Indo–Pacific, greater along the west American shelf where large continents border a large ocean, and greatest in the Atlantic where large continents face a small ocean.

10.3.3   Zoogeographical barriers

New species tend gradually to extend their ranges beyond the original location. Many species that evidently arose in the Indo–Malayan region, for example, now occur throughout the vast tropical Indo–West Pacific area. Physical features of the oceans and continents, such as unfavourable currents, large stretches of cold, deep water or shallow, warm water, and land barriers, may impede range extensions, with important consequences to the species diversity of shelf faunas.

The East Pacific Barrier (Fig. 10.7) is a wide stretch of deep water between Polynesia and America that has greatly restricted the spread of tropical, Indo–West Pacific, shallow-water species to the coasts of tropical America. Only about 6% of tropical, Indo–West Pacific, shore fish have managed to cross the East

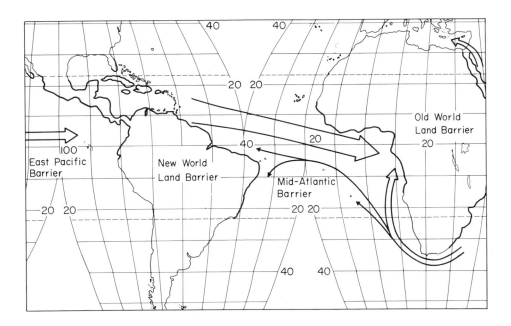

Fig. 10.7 Zoogeographical barriers separating the tropical shelf regions. The arrows indicate the direction and approximate relative amount of colonization that has occurred. (After Briggs 1974.)

Pacific Barrier, as has a similarly small proportion of shallow-water invertebrates. Consequently, the tropical East Pacific shelf fauna is depauperated by the existence of the East Pacific Barrier. Due to the prevailing currents, no species originating in the eastern Pacific have managed to extend westwards across the East Pacific Barrier.

Other barriers with similar zoogeographical effects (Fig. 10.7) include the New World Land Barrier that prevents movement of tropical marine species between the eastern Pacific and western Atlantic, the Mid-Atlantic Barrier, a broad stretch of deep water that separates tropical western and eastern shelf faunas, and the Old World Land Barrier that separates the tropical eastern Atlantic and Mediterranean from the Red Sea and tropical Indian Ocean.

Warm, shallow, tropical shelf waters isolate the shallow, benthic faunas of northern and southern temperate regions. A number of species however such as the barnacle *Balanus balanus* and the colonial tunicate *Botryllus schlosseri* (see Fig. 9.16a) are able to extend across the equator by living in deeper, cooler water, a phenomenon known as equatorial submergence.

10.3.4 Deep-sea benthos

Deep-sea, benthic, faunal diversity is extraordinarily high (Chapter 7). Although the physical conditions are extreme compared with those of more familiar shallow habitats, they are stable, as is the food supply. Dead planktonic organisms and faecal pellets sink very slowly and take so long to descend to the sea bed that pulses of primary production at the surface become smoothed out. Large particles, such as dead vertebrates, are quickly consumed by scavengers, as demonstrated by the rapid exploitation of bait lowered to the sea bed. Scavengers, mainly fish, disperse the organic matter as faeces, thereby evening out the spatial distribution of food for deposit feeders.

The predictability of the deep-sea environment would therefore seem conducive to specialization. However, the problem remains of explaining how competitive exclusions are prevented in such an apparently homogeneous habitat and how gene pools diverge in the apparent absence of barriers to gene flow. Dayton and Hessler (1972) propose that generalist 'croppers' keep prey population densities sufficiently low to prevent competitive exclusions (Chapter 7). But would not the evolutionary arms race between predators and prey tend to offset this in such a stable environment? Hessler and Sanders (1967) suggest that the long geological history of stability in the deep sea has allowed the evolution of numerous species with narrow niches, but they do not indicate how the resources are partitioned or how genetic divergence and reproductive isolation take place. It has been suggested that many deep-sea species have not evolved from common ancestors *in situ* but are derived from ancestral immigrants from shallow waters. Comparative morphological studies of deep-sea isopods, however, indicate a long history of *in situ* evolution. The explanation of speciation and faunal diversity in the deep sea therefore remains an important problem, for the deep-sea environment accounts for a very large proportion of the biosphere.

10.3.5 Latitudinal zonation

Just as species are vertically zoned on shores in response to the terrestrial–marine environmental gradient (section 5.1.1), so they also tend to be horizontally zoned on a geographical scale in response to a complex gradient associated with latitudinal changes in climate. The latitudinal climatic gradient is particularly pronounced on the east coast of North America. The south-flowing, cold Labrador Current and the north-flowing, warm Florida

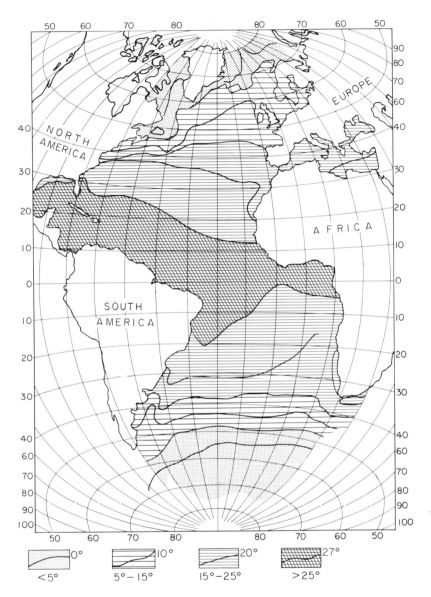

Fig. 10.8 Mean annual temperature isotherms of the surface water of the Atlantic, showing the compressed temperature gradient on the east coast of North America. (After Ekman 1967.)

Current cause a very rapid change from arctic to tropical temperature regimes within 20° of latitude (Fig. 10.8). Not surprisingly, the latitudinal zonation of many eastern North American intertidal species is correlated with critical boundaries where

271

temperatures become too extreme for reproduction or adult survival.

Along the western coast of North America the climatic gradient is much more gradual and factors other than temperature become important in determining latitudinal, zonal boundaries. Behrens-Yamada (1977) identified factors limiting the latitudinal distribution of two species of periwinkle by transplanting the snails beyond their normal ranges. The southern limit of *Littorina sitkana* and the northern limit of *L. planaxis* occur at about latitude 43°N. *L. planaxis* transplanted further north survived and reproduced for four years, but the potential northerly range of this species is curtailed by south-flowing currents that prevent the northward spread of the pelagic larvae. *L. sitkana* transplanted south of its natural range succumbed both to desiccation, because of a lack of damp microhabitats, and to predation by crabs, which are more abundant in the high intertidal zone to the south.

10.3.6 Plate tectonics and provincialization

Waves of extinction and taxonomic diversification are prominent features in the fossil record. Sometimes extinction coincided with a reduction in taxonomic diversity (diversity-dependent extinction), suggesting that the carrying capacity of the biosphere had somehow declined, but at other times waves of extinction did not affect taxonomic diversity (diversity-independent extinction), evidently because lost lineages were replaced quickly by new ones. The most likely cause of diversity-independent extinction is a change in climate that causes unadaptable species to die out but does not affect the carrying capacity of the biosphere, so that taxa soon evolve to utilize vacated niches.

A density-dependent wave of extinction occurred in the Mid to Late Permian (just over 200 million years ago), reducing the diversity of most taxa to their lowest levels since the Cambrian when fossils first became numerous. The causes of the Permian crash in taxonomic diversity remain in debate, but the surge of rediversification during the Mesozoic–Cenozoic period (180–120 million years ago) is much better understood.

Prior to the Mesozoic period the continents were grouped together, surrounded by the huge Pacific ocean and partly divided by the Tethys Sea (Fig. 10.9). During the Mesozoic period the continents began to separate owing to sea-floor spreading and this must have altered the climate considerably (for an account of plate tectonics and its biogeographical implications see Valentine

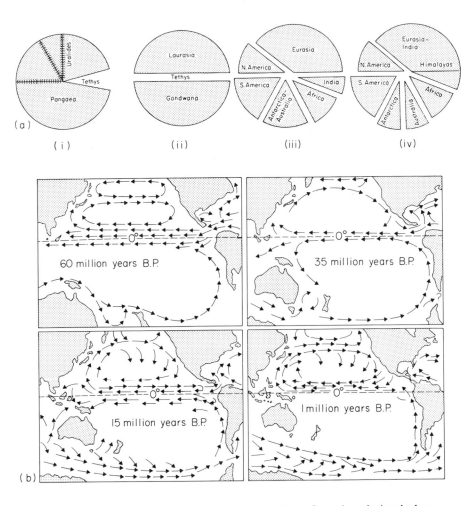

Fig. 10.9 (a) Inferred continental and oceanic configurations during the last 200–300 million years. (i) In the Permean, Pangea is partly intersected by the Tethys Sea. (ii) In the Early Mesozoic, Laurasia and Gondwana are separated by the Tethys Sea. (iii) At the end of the Cretaceous, Gondwana is highly fragmented. (iv) At the present, India is joined to Eurasia. (After Valentine 1973.) (b) Changing configuration of continents in the Pacific region during the last 60 million years and the associated changes in surface currents. (After van Andel 1979.)

(1973) and Gray and Boucot (1979)). The huge continuous area of the pre-Mesozoic supercontinent would have intensified temperature fluctuations, producing a more continental world climate (section 10.3.2). Separation of the continents had several important effects. First, the increased coastline per unit area of land ameliorated the world climate, making continental shelf areas more stable and therefore more conducive to the evolution of species with specialized niches (section 10.3.1.2). Second, separation of the continents promoted faunistic and floristic isolation, increasing the opportunities for allopatric evolution. Consequently, biotas of geographically separate shelves developed different taxonomic compositions, or in biogeographical terminology, developed provincial differences. During the early Jurassic there seems to have been no provinciality at all, since ammonite and bivalve taxa were cosmopolitan. By the Middle and Late Jurassic there were two, well-defined faunal realms and today over 30 marine provinces are recognized. Third, the changing positions of the continents altered the pattern of oceanic circulation so as to intensify the equatorial–polar temperature gradient. Owing to the isolation of the Arctic Ocean by continental topography and of the Antarctic seas by the Circumpolar Current, polar seas at present are as cold as they can ever be. The pronounced, latitudinal, temperature gradient led to latitudinal zonation (section 10.3.4) and increased provinciality. The glacial cycle, which may have started in association with changing continental configuration and oceanic circulation, has not caused massive extinctions. This is because the predominantly north–south continental orientation allows species to move gradually with the latitudinal advance and retreat of cooler conditions.

10.3.7 The effects of man

Current theories of plate tectonics and evolutionary ecology could not be expected to predict the small but significant effects that man himself has had on contemporary biogeography. Man's technology has resulted in the passive transportation of organisms throughout the world and even the opening up of new dispersal routes between previously isolated seas.

Organisms are dispersed attached to the hulls of ships or to commodities such as oysters. Most hitch-hikers will be killed in transit or will fail to establish themselves in foreign localities because conditions are unsuitable for their growth or reproduction, or because there are too few simultaneous arrivals to form a viable breeding group (propagule). Opportunistic (r-selected) species (section 9.1), with their high potential population growth rates and ability to flourish from small initial densities, are the most likely organisms to spread by hitch-hiking. Examples include the Australasian barnacle *Elminius modestus* which was first noticed in Chichester Harbour, England, in 1945 and now thrives in moderately sheltered localities throughout Britain. Presumably *Elminius* was given the opportunity to breed and recruit young while ships were laid up during the Second World War. In 1890 a shipment of American oysters brought the slipper limpet *Crepidula fornicata* to Essex, since when the stowaway has become firmly established in the oyster beds of south-west Britain. *Littorina littorea* was apparently taken from Europe to North America several hundred years ago and is now a dominant species from Newfoundland to New England.

Successful immigrants can make an impressive ecological impact. In the early nineteenth century the cord grass *Spartina alterniflora* was introduced from America to Southampton Water where it hybridized with the European species *S. maritima*, forming a sterile hybrid *S. townsendii*. This hybrid subsequently became polyploid, forming the sexual species *Spartina anglica*, which is so vigorous that it has spread throughout much of the British Isles, forming dense stands where once there were open mud flats. Animals such as *Scrobicularia plana* and *Macoma balthica* disappear as *S. anglica* monopolizes the substratum, but other species such as *Littorina rudis* and *Anurida maritima* move in.

As well as carrying organisms about with him, man has assisted their dispersal by connecting the Red Sea to the Mediterranean. Organisms began to disperse along the Suez Canal immediately after its opening in 1869. Currents flow from the Red Sea to the Mediterranean for ten months of the year so that most invertebrate migration has been in this direction. At least 140 animal species, including fish, decapod crustaceans, molluscs, polychaetes, ascidians and sponges have reached the Mediterranean via the Suez Canal.

The Panama Canal connecting the Pacific to the Atlantic has a salinity of less than one per thousand and therefore can only be negotiated by a few euryhaline species such as the blue crab *Callinectes sapidus*. The proposal for a sea-level canal, if implemented, would probably have important biogeographical consequences. The eastern and western Central American shelf faunas contain approximately 8000 and 6000 species respectively. Bringing these species into contact could cause the extinction of many of them.

Finally, man has severely depleted many of the populations he exploits. Stellar's sea-cow (p. 148) and the great auk were exterminated, and the large whales seem dangerously close to a similar fate. Ecological consequences of such overindulgences are not yet predictable, but could be far-reaching. It has been suggested that decimation of the trumpet triton, *Charonia tritonis*, on the Great Barrier Reef is at least partly responsible for the outbreaks of

Chapter 10

Acanthaster planci (p. 176). Decimation of whales has made more planktonic production available to other consumers (p. 293), but technology is being developed to exploit the prodigious productivity of antarctic krill, the very key to the evolution and continued existence of the larger baleen whales.

Will the future fossil record tell the tale of man's profligacy? Human impact on the marine biota is, at present, miniscule compared with biogeographical events that have occurred in the geological past. If man does have a major biogeographical influence on marine life, it will probably be through the effects of pollution or atomic war. We can only hope that this is not as likely as the coming of the next glacial period.

Useful texts on speciation, biogeography, and plate tectonics in the marine context include Battaglia and Beardmore (1978), Briggs (1974), Ekman (1967), Gray and Boucot (1976) and Valentine (1979).

11 The Marine Ecosystem as a Functional Whole

By now the broad outlines of the interrelationships between different marine organisms and different marine habitats should be fairly clear. It is the purpose of this chapter to quantify some of the fluxes between the different compartments in the overall system and to draw together the various threads introduced in the earlier chapters. First, however, it may be helpful to summarize the broad outlines referred to above.

11.1 The general nature of interactions within the marine ecosystem

Excluding the centres of primary chemosynthesis on the sea bed, the initial production of organic compounds in the ocean is confined to the shallow, photic zone. In this surface water layer and in those littoral areas bathed by it, carbon dioxide is fixed by photosynthesis and nutrients are incorporated into living tissues. The dominant organisms achieving this primary production are the minute protistan phytoplankton together, in the littoral zone, with the related benthic species, some of which, in the form of seaweeds, attain macroscopic size. Also in the littoral zone are semi-aquatic stands of plants of terrestrial affinity which are highly productive but which are relatively little grazed by consuming species. Neither are the larger seaweeds consumed directly to any appreciable extent, except when at the sporeling stage, although their finer, filamentous epiphytes can be consumed as efficiently as the free-living protists. The uni- or non-cellular and smaller multicellular photosynthetic protists therefore provide a direct source of food for herbivorous species whilst the production of the coarser, macrophytic seaweeds and of the semi-terrestrial tracheophytes enters marine food webs largely as detritus. Together their biomass has been estimated to total some 4×10^9 t (dry weight) of organic matter, and this produces annually some 55×10^9 t (dry weight), or about fourteen times the standing crop (Table 11.1). (Compare values on the continents of 1837×10^9 t and 115×10^9 t respectively.)

Most of this photosynthetic production is consumed in the same surface waters: the phytoplankton by herbivorous

Table 11.1 World marine photosynthetic biomass and productivity.

	Biomass 10^9 t	Productivity 10^9 t per year
Oceanic	1.0	41.5
Neritic and upwelling	0.3	9.8
Littoral	2.6	3.7

277

zooplankton and, in shallow waters, by the suspension-feeding benthos, and the littoral benthic algae by grazing herbivores (gastropod molluscs, etc.). A large, but largely unknown, proportion of the gross primary production is also leaked into the surrounding water in the form of dissolved organic matter. This enters a parallel food chain based on bacteria and heterotrophic protists which are consumed by other zooplanktonic species. Assimilation efficiency of the zooplankton and of the shallow-water benthos is often low and their faecal pellets contain much material of potential food value. These faeces, together with detrital substances derived from the fringing tracheophytes and seaweeds (and in some areas from unconsumed phytoplankton), form the substrate for bacterial growth and, ultimately, for bacterial-dependent populations of heterotrophic protists and small metazoans. As these aggregations sink through the water, so they are consumed by members of the zooplankton which digest the living microbiota and microfauna, until eventually the recycled aggregations reach the sea bed. There they support a deposit- and suspension-feeding benthos which, in regions below the photic zone, is ultimately similarly dependent on bacterial production. In some areas of sea bed, genuinely primarily producing bacteria can fix carbon dioxide by chemosynthesis and these too will pass organic matter into the benthic food web, as will those bacteria which chemosynthetically can regenerate organic matter using the end products of previous photosynthetic activity.

Herbivorous and detritivorous planktonic organisms are consumed by the carnivorous zooplankton, and equivalent benthic carnivores take the deposit and suspension feeders. All these planktonic and benthic consumers then form the food of the nekton, which occupy the uppermost portions of the marine food web over all but the deepest regions of the abyssal trenches. The large majority of the nekton are carnivores although a limited number of species can consume the primary producers and even fewer appear able to take the detrital/bacterial aggregates.

As a consequence of primary production being a surface-water phenomenon and because shallow-water zones are the most productive regions of the sea per unit area, secondary production declines with both distance away from the coast and depth in the sea. Where deep waters lie immediately adjacent to the coast, however, detrital bonanzas may occur at great depths, and debris from coastal tracheophytes is found over large areas of the abyssal plain; nevertheless, the abyssal benthos is the slowest growing, longest lived and least productive of all marine systems. Relative food shortage places a selective premium on efficiency of

zooplankton and, in shallow waters, by the suspension-feeding benthos, and the littoral benthic algae by grazing herbivores the deep sea, and least efficient in areas of upwelling, in estuaries and in shallow waters generally.

Total animal biomass in the sea has been estimated at 1000×10^6 t (dry weight), a comparable value to that occurring on the continents. However, this biomass probably produces more than three times more organic matter per year than does the terrestrial/freshwater fauna; the marine biomass produces 3025×10^6 t (dry weight) as compared with the continental production of 900×10^6 t (dry weight). Thus, although marine primary production is much lower than that on land, marine secondary production is much greater (primary to secondary production ratios of $1:0.008$ on land and $1:0.06$ in the sea). This is a result of the greater digestibility of marine primary production, of the greater efficiency of trophic relationships in the sea, and, in part, of the poikilothermic nature of marine consumers (less of the food energy has to be devoted to fueling a temperature differential between body and environment). Man currently harvests 60×10^6 t (wet weight) of marine production, almost entirely in the form of the carnivorous nekton.

So far in this synopsis, we have been following the passage of food materials from their fixation to the top consumers, but essential elements such as nitrogen are present in finite amount and unless these are cycled through the ecosystem back to the primary producers, productivity will cease. Supply of nutrients is probably the most frequent rate-limiting process in the sea, both to the photosynthesizers and to the bacteria. Nutrients are released from their organic binding by the animal consumers during their metabolism; ammonia and phosphates are then excreted or diffuse back to the water mass from where they may be withdrawn again by bacteria, algae, etc. Nevertheless, some nutrients will be incorporated in the detrital and faecal materials which sink towards the sea bed, so that although most nutrient regeneration probably occurs *in situ* in the photic zone, there is a continual draining of these essential substances from the surface waters.

In shallow regions, tidal currents and wind-induced mixing will suffice to reinject those nutrients remineralized by benthic animals into the photic zone (if indeed the photic zone does not extend all the way to the sea bed), but nutrient loss by sinking towards the bottom is particularly important in waters overlying a thermocline which separates permanently the photic zone from deeper waters and from the benthos. Nutrients regenerated below the thermocline will tend to remain at depth, except where deep water (i.e. down to 600 m) upwells to the surface. Hence nutrient cycling will be most unconstrained in shallow waters and in zones of upwelling, and, all things being equal, these areas will therefore be maximally productive, although often still nutrient-limited. In waters with small stocks of nutrients, high productivity can only be maintained by a very rapid recycling (aided, in coral reefs, by symbiotic associations) or by the presence of moneran species capable of fixing atmospheric nitrogen.

Some nutrients are incorporated into organic substances which are highly resistant to decomposition. Although these may go into solution, their half-life as dissolved organics is likely to be very long and although a huge pool of these substances is present in the ocean it is likely to be inert. Other dissolved organic materials released by both algae and the animal consumers may be cycled, but much more slowly than the inorganic nutrients required by most photosynthetic organisms. Productivity may then be dependent on the rate of remineralization of this second pool. Standing stocks of nutrients are often of very little interest in marine ecology; it is the flux rates which govern photosynthetic production. At any one time, the finite pool of, say, nitrogen will be divided between living tissues, inorganic forms dissolved in the water, dissolved organically bound nitrogen, and particulate organic nitrogen in detritus and faeces; and exchanges between these compartments may be fast or slow. Unfortunately, we know much more about the size of these pools than of their dynamic interaction.

This then is a summary of the patterns with which carbon (or energy) flows through the marine ecosystem and in which nutrients must cycle (see Fig. 11.1), and we will now attempt to quantify some of the fluxes.

11.2 Energy flow

There are so many unknowns in the precise nature of marine food webs that it takes a brave person to propose a budget for any section of this ecosystem, let alone for the whole world ocean. Budgets for single-species populations are not uncommon and an increasing number are appearing for localized regions of specific habitat type; the numbers of those treating larger areas, however, are very few. One particularly instructive example of the latter is that of Pomeroy (1979) for neritic seas and the continental shelf, and this we shall now consider in some detail.

Pomeroy investigated various potential patterns of energy flow

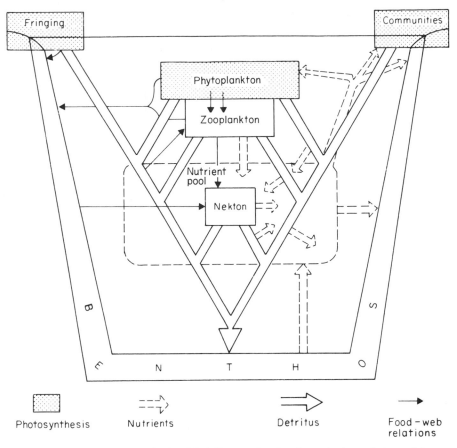

Fig. 11.1 A greatly simplified diagram showing the major interrelations of the four semi-isolated subsystems into which all aquatic systems, including the ocean, can be divided. (After Barnes 1980b.)

through the food web shown in Fig. 11.2 which represents a generalized pelagic and benthic system over any continental shelf. This web differs from that outlined in the previous section only in that microzooplanktonic and meiobenthic organisms are omitted (because although they are undoubtedly important their exact significance is still unknown), and in that two trophic categories are subdivided. The herbivorous zooplankton have been divided into 'grazers' and 'filterers'. Grazers include those species that seize relatively large phytoplankton cells individually or capture them in a filtration system of comparatively large mesh; e.g. crustaceans using the setae on their mouthparts or legs as a sieving screen. The filterers, on the other hand, feed on the smaller, nanoplanktonic algae and on bacteria which they can retain on mucous filters of extremely small pore size; the pelagic tunicate appendicularians, for example, can filter from the water material of less than 1 μm in size. The second, subdivided, trophic category comprises the detrital products deriving largely from the phytoplankton: algal material in faecal pellets is distinguished

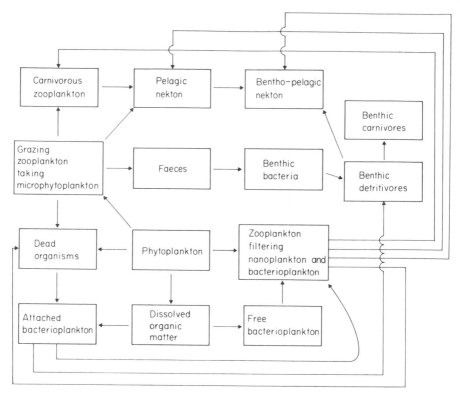

Fig. 11.2 A food web for the neritic–shelf marine system with only the energetically more important links shown. Omitted, for example, are imports from the littoral system and interchange with the oceanic–abyssal region; further, in reality all organisms will contribute to the 'dead organisms' category, and some benthic carnivores will also be taken by bentho-pelagic nekton. (After Pomeroy 1979.) See also Figs 11.3, 11.4 and 11.5.

from the dead remains of algae killed, but not eaten, by grazers. It has been estimated that 25% of phytoplankton production is broken as a result of unsuccessful feeding attempts by grazing consumers, the proportion perhaps increasing to 50% in areas or at times of relative food abundance. Dead and dying algal cells are therefore the dominant element of the 'dead organisms' compartment in Fig. 11.2.

For simplicity, Pomeroy assumed that neritic seas did not receive any import of detrital materials from the littoral zone, and that there was no transfer of material between the oceanic region and coastal waters; neither were benthic suspension and deposit feeders distinguished. The food web was fueled by an input of 1000 kcal m^{-2} per year of phytoplanktonic productivity (10 kcal $\equiv$ 1 g C fixed). This is within the range given on p. 38 for neritic production, although only 60% of the average value suggested there. Nevertheless, it does have the advantage of permitting proportions of the total productivity flowing through different pathways to be seen at a glance. This input was then influenced by two parallel series of variables: (a) the proportion consumed when alive or deriving from differently sized phytoplankton species; and (b) different levels of efficiency of energy transfer from consumed to consumer. The first set of variables comprised three alternative states: (1) transfer of energy from microphytoplankton (e.g. diatoms) to grazing herbivores dominates the energy flow; (2) mucus-net filterers are the dominant consumers taking abundant nanoplanktonic species or bacteria; or (3) detritus dominates the flow pattern, as, for example, during the late stages of a phytoplankton bloom when most algae die rather than are consumed alive. The efficiency of energy transfer to bacterial tissue was set at 50% throughout the analysis, and that to nektonic and benthic carnivores remained constant at 10%. But two alternative sets of efficiencies were applied to the other links in the food web: a generally low series of mucus-net filterers (17%), grazing zooplankton (19%), carnivorous plankton (13%), and benthic detritivores (33%); and a higher set of 30% for all these four groups.

Combining these two sets of variables whilst maintaining a fixed input of 1000 kcal m^{-2} a year yields six different energy-flow patterns, each with individual values of steady-state biomass in the various interacting compartments, dependent on energy receipt and loss (Figs. 11.3–11.5). Each of the six is realistic in the sense that the division of the basic energy supply between the four compartments radiating from the phytoplankton (grazers, filterers, dissolved organics, and dead tissue) is in accord with known field measurements in neritic seas. What then are the differences between the six patterns? Comparing Figs. 11.3a, 11.4a and 11.5a, it can be seen that when most production is in the form of the larger members of the phytoplankton, not only is the grazing zooplankton abundant but so also are the benthic consumers which feed on the rain of faecal pellets (Fig. 11.3a). In the majority of the crustacean grazers, faecal pellets are encased in a chitinous tube and they remain intact for about two days, meanwhile sinking rapidly to the sea bed. Diverting algal consumption to those zooplankton which use mucus filters decreases the input to the benthos (Fig. 11.4a) but increases the quantity of dead organic matter in the water. The faeces of these species are more diffuse, remain for longer in the water column and are more readily colonized by bacteria, whilst in addition their dead or discarded tests ('houses') add substantially to the suspended organic load. Nektonic production does not change appreciably as a result of this alteration in the dominant zooplankton types. Finally, when the phytoplankton cells die at the end of a bloom (Fig. 11.5a), nektonic production decreases and benthic

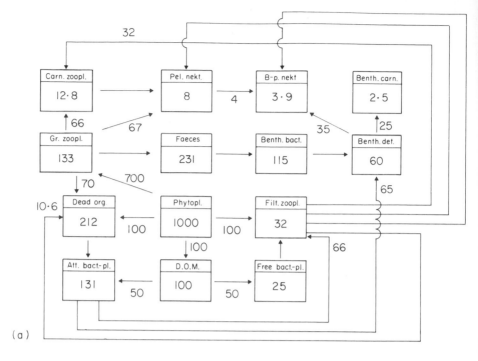

(a)

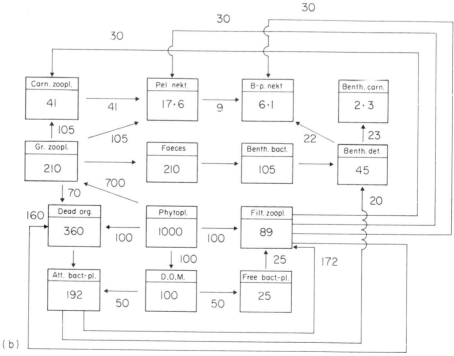

(b)

Fig. 11.3 Energy flow through the food web shown in Fig. 11.2 under circumstances in which the grazing zooplankton consuming microphytoplankton receive all but a minor portion of the total phytoplanktonic production (set arbitrally at 1000 kcal m^{-2} per year). The values inside the compartments are the average biomass in kcal m^{-2}; the values alongside the arrows are the flow rates in kcal m^{-2} per year. (a) With relatively low efficiencies of energy transfer through intermediate food links; (b) with higher efficiencies of flow through those links (see text). (After Pomeroy 1979.)

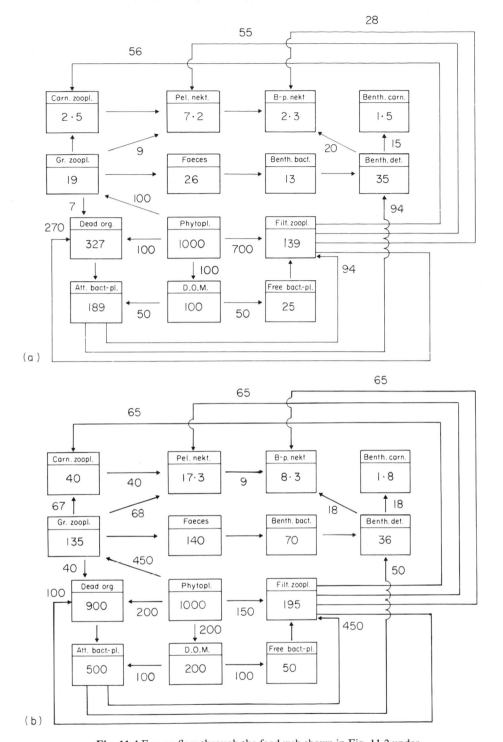

Fig. 11.4 Energy flow through the food web shown in Fig. 11.2 under circumstances in which the filtering zooplankton (e.g. species using mucus nets) are relatively important, feeding either on nanoplankton or on bacteria. Total phytoplanktonic production is set at 1000 kcal m^{-2} per year. The values inside the compartments are the average biomass in kcal m^{-2}; those alongside the arrows are the flow rates in kcal m^{-2} per year. (a) With relatively low efficiencies of energy transfer through intermediate food links; (b) with higher efficiencies of flow through those links (see text). (After Pomeroy 1979.)

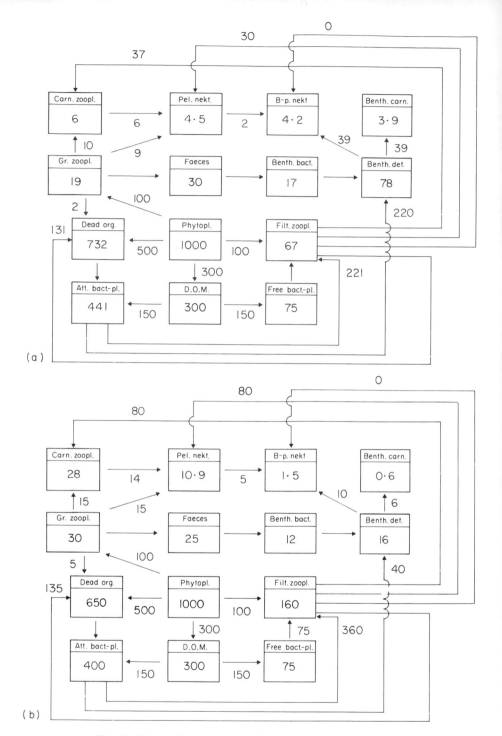

Fig. 11.5 Energy flow through the food web shown in Fig. 11.2 under circumstances in which dead phytoplankton cells are abundant and much dissolved organic matter is released into the water (i.e. circumstances typical of the late stage of a bloom). Total phytoplanktonic production is set at 1000 kcal m^{-2} per year. The values inside the compartments are the average biomass in kcal m^{-2}; those alongside the arrows are flow rates in kcal m^{-2} per year. (a) With relatively low efficiencies of energy transfer through intermediate food links; (b) with higher efficiencies of flow through those links (see text). (After Pomeroy 1979.)

production is maximal, largely as a consequence of bacterial abundance.

If one compares the productivity of the terminal links in these three flow diagrams (and the biomass which can be supported) with actual field information, they are all relatively low, and would be even lower if microzooplanktonic and meiobenthic compartments had been included in the model. This can be corrected either by boosting the value assigned to primary production (which as we have noted is at the low end of the spectrum) or by increasing the efficiency of energy transfer. There is some evidence that phytoplanktonic productivity as measured by the C^{14} method (including all those values given on pp. 37–9) may be an underestimate of the true rate (Peterson 1980), especially where the dominant algae are nanoplankton. The matter is still unresolved, but it has been suggested that planktonic primary production is really an order of magnitude larger than the values generally accepted and presented here. Alternatively, the efficiency of energy transfer may be much higher than the efficiencies used to produce Figs. 11.3a, 11.4a and 11.5a. Much argument remains as to the most appropriate efficiencies to use (see pp. 63–5) but Pomeroy, at least, considers that the values given in Table 2.3 provide a more realistic estimate of the transfer of energy through marine food webs.

Using the higher set of efficiencies in place of those of around 20 % (i.e. an increase of 50 % in the efficiency of energy transfer along intermediate links) does result in increased biomass and productivity of the nekton, but not of the benthos (Figs. 11.3b, 11.4b, 11.5b). Indeed, under late bloom conditions (Figs. 11.5b) the benthos is greatly reduced (cf. Fig. 11.3b). Separation of suspension and deposit feeders in the model, and diversion of energy through the suspension feeders, would probably result in an increase in the first, though not in the second: as we have seen, short-term cycles do occur in the benthos and suspension feeders may increase whilst deposit feeders are declining. In a sense, this is a mere technicality. The models of Figs. 11.3–11.5 are undoubtedly incomplete, but they are the most realistic yet devised. At a minimum, they indicate that our estimates of either primary production or of ecological efficiencies (or of both) are likely to be too low. At a maximum, when refined and modified by future information, they will be able not only to describe quantitatively the neritic/shelf food web—and ultimately that of the larger oceanic region as well—but they can also be used to predict the effects of perturbation of these systems. By altering the patterns in which energy flows, for example by decreasing greatly the importance of some links, the effects on the remainder of the system can be gauged by a relatively simple laboratory simulation; an operation impossible to perform in the field. It is important to remember, however, that any such model ecosystem is only as good as the assumptions on which it is based and as realistic as the data fed into it.

11.3 Nutrient cycles

As stated earlier in this chapter, as yet we are ignorant of the magnitude of fluxes of nutrients through any part of the marine ecosystem, although we do know the sizes of some of the standing stocks or pools. Without knowledge of rates of turnover, however, information on concentrations per unit volume in different compartments is of very little use. The filling of this gap probably represents the most urgent problem in marine ecology, and the most difficult. The ocean is less self-contained with respect to nutrient cycles than in almost any other regard. The seas form one compartment in a global ocean–atmosphere–land system and interchange between all three occurs. Interchange has also increased greatly in recent years as man has mined terrestrial sources of nitrates and phosphates, has spread them on the land, and has thereby permitted rainfall and river flow to wash them into the sea. Woodwell (1980) suggested that man may now be responsible for a throughput equivalent to 30 % or more of the natural global flux of nitrogen. Another complication is that nutrients may cycle both chemically and spatially.

Chemical cycling is relatively straightforward and we can illustrate it by considering nitrogen. By far the largest source of nitrogen is the gaseous form, molecular N_2, present freely in the atmosphere and dissolved in natural waters. This, however, is unavailable to all organisms except various of the moneran blue-green algae and bacteria. Although some physical processes (lightning, vulcanism, etc.) can fix molecular nitrogen and man is contributing increasing quantities of fixed nitrogen to the biosphere, it is these micro-organisms which are mainly responsible for making nitrogen available to other organisms (and for releasing it back to the atmosphere under anoxic conditions). The fixed nitrogen then cycles chemically through the ecosystem as shown in Fig. 11.6.

Not all parts of such cycles are distributed equally through all parts of the marine ecosystem: some, such as remineralization of refractory particulate organics and denitrification of nitrate to ammonia, are located typically in the sea bed; others,

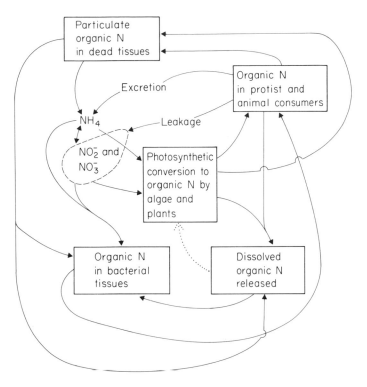

Fig. 11.6 The chemical cycling of fixed nitrogen.

photosynthetic incorporation of inorganic nitrogen into organic forms for example, are confined to the surface waters. To illustrate the gross spatial cycling of nutrients, we can consider the other major photosynthetic requirement—phosphorus (Fig. 11.7). Over the ocean as a whole, the dominant cycle is clearly incorporation into organic substances, descent into the ocean depths and reinjection into the surface by upwelling: 95 % of the upwelled inorganic phosphorus descends each year in organic form. Such gross analysis of the cycle, however, can be highly misleading. In practice, each phosphorus atom in the photic zone may pass through many cycles of fixation → consumption → excretion → fixation before it finally descends into the aphotic depths. We saw on p. 66, for example, that the turnover time of phosphate in temperate seas is only 1.5 days and on p. 191 that at least 75 % of the phosphorus pool on a coral reef is turned over each day. Therefore between injection into the surface waters and final descent out of the photic zone, each phosphorus (or nitrogen) atom will have passed through very many organisms and will have been

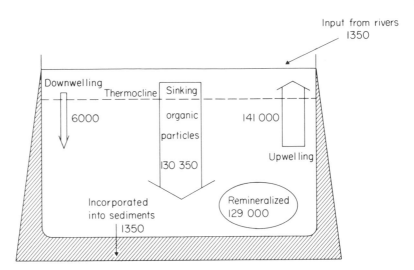

Fig. 11.7 The gross annual flux of phosphorus (10^6 kg per year) through the oceans. (From data in Wright 1977–78.)

recycled rapidly. To date, we have a limited series of spot measurements or estimates of the rapidity of recycling; a much larger series is required to support generalizations and the collection of this information is still a matter for the future.

12 Human Exploitation and Interference

The main purpose of this book is to describe the natural patterns and processes of the marine ecosystem, but man's influence on all natural systems is increasing year by year and hence it is necessary to add a final chapter on the effect of our own species on marine ecology. Here we will investigate three aspects of man's utilization of the sea: (1) exploitation of marine organisms for food and other purposes; (2) use of the sea as a dumping ground for unwanted wastes; and (3) conversion of shallow marine environments into building or agricultural land, freshwater reservoirs, and other non-marine habitats. These are all very large subjects—and man affects the sea in many more ways than the three specified above—hence our treatment must be selective and superficial. More detailed information may be found in Hood (1971), Goldberg (1972), Ketchum (1972), Nelson-Smith (1972), Cushing (1975), Cushing and Walsh (1976), Barnes (1977a), Knights and Phillips (1979) and Odum (1980).

12.1 Fisheries

The current harvest of marine organisms is shown in Tables 12.1 and 12.2. It is clearly a yield dominated by fish, caught by essentially primitive hunting techniques (although with an arsenal of modern technological aids). The harvest differs markedly from that taken from the land, both in size and in nature. In terms of human food, the land and fresh waters provide 99.2% of our intake and the sea, therefore, only 0.8%. Fish and fish products are an insignificant component of the human diet in general. In only two contexts are they important: luxury foods (salmon, tuna, prawns, etc) which command a high price; and staple foods in relatively poor countries, especially as a source of protein. In Bangladesh, for example, 80% of the animal protein consumed is in the form of fish. A second major difference between terrestrial and marine harvests is the position in the food web occupied by the food items. Plants comprise 94% of the yield from the land, and herbivores form the remaining 6%. Most marine fish are predators and hence 80% of the marine yield is in the form of carnivores; photosynthetic organisms comprise a minute proportion. In part this simply reflects the ease of harvesting and processing the respective foods, and also human food preferences; but the greater quantity of secondary production in the sea in respect of unit primary production (p. 279) may also be relevant. At least it is more efficient to crop marine carnivores than it would be to do so on land.

Be that as it may, man does not use marine nektonic production

Chapter 12

Table 12.1 Commercial landings of marine organisms in 1977 (in $t \times 10^6$). (From data in FAO 1978.)

Seaweeds	1.5
Benthic crustaceans	2.2
Pelagic crustaceans	0.1
Benthic molluscs	2.9
Pelagic molluscs	1.2
Other invertebrates	0.1
Diadromous fish*	1.6
Marine fish	54.0
Turtles	0.007
Mammals†	0.5
Total	64

* Fish that migrate between fresh water and the sea, e.g. salmon and eels.
† 15 000 individual medium/large whales and 332 000 pinnepeds plus 3500 t sea-cows, small whales, etc.

Table 12.2 Landings of different groups of marine fish comprising the 1977 commercial catch* (in $t \times 10^6$). (From data in FAO 1978.)

Herring, anchovy, etc.	13.0
Gadoids (cod, haddock, etc.)	10.7
Jacks, sauries, etc.	8.7
Redfish, sea-bass, conger, etc.	5.1
Mackerel, etc.	3.6
Tuna, etc.	2.3
Flat-fish (plaice, halibut, etc.)	1.1
Elasmobranchs (sharks, etc.)	0.6
Miscellaneous others	8.9
(Diadromous species	1.6)
Total	55.6

* The landings resulting from 'sport fishing' are excluded from the commercial catch.

efficiently. Some 30–40% of the world fish catch is not consumed directly, i.e. in the form of fish, but is converted to oil or fish-meal for use as agricultural fertilizer or as animal feed (being fed, for example, to broiler chickens). Many people, therefore, only benefit from marine productivity after it has passed through another (terrestrial) food chain with all the losses which that entails. Fisheries with a fish-meal end-product are characteristic of upwelling regions. Until its recent collapse, the largest single fishery in the world, which in its heyday of 1967–1970 comprised 20% of the world's catch and made Peru the premier fishing nation, was based on converting the anchoveta, *Engraulis ringens*, into fish-meal. (The revenue from fish-meal accounted for 33% of Peru's export earnings.)

The current world harvest of marine organisms has been fairly

292

stable at 60×10^6 t (fresh weight) for the last ten years. What is the potential for increasing this yield? Some large estimates of potential catch have been produced, but these are based on all known fish populations in the sea, including the small fish present at very low population densities in the deep sea. To catch the 500 t of fish required to make an ocean-going trawler profitable, it would, for example, be necessary to catch all the deep-sea fish present in a volume of water of 10 km^3. This is completely out of the question at the moment. If calculations of potential catch are restricted to the general types of fish already being caught, the estimate reduces to a maximum value of not much over 100 $\times 10^6$ t —a maximum potential catch only two-thirds as much again as is currently being taken. This could be achieved by exploiting new fishing grounds (such as the Indian Ocean), rather than by fishing more intensively existing grounds, as many of these are already overexploited (see below). To this estimate of annual fish catch can be added 10×10^6 t of squid and, possibly, up to 100×10^6 t of Antarctic krill, *Euphausia superba*, 'released' by the human reduction in numbers of one of their main predators, the baleen whales (p. 299). These additional yields, however, are still speculative.

The estimates quoted above are the 'maximum sustainable yields' of the organisms concerned, the important word being 'sustainable'. If one does not put much effort into trying to catch something, the catch will be small; larger catches would result from the application of more effort. Equally clearly, however, if by making too great an effort one catches too many of the organism in question, the ability of its population to make good the losses will be impaired and future catches will decline. Somewhere between these two states will be an effort yielding the maximum sustainable catch (Fig. 12.1), and much applied ecological research is devoted to ascertaining the value of this yield for each of the various exploited fish stocks. (Of course, calculation of the 'MSY' is not nearly so straightforward a matter as might be implied by the simplified argument above: variations in recruitment, natural mortality and many other factors have to be taken into consideration.)

Yield, however, is a product of two separate variables: numbers of fish and average size (p. 213). Some fish, flat-fish for example, continue to grow after reaching maturity and can double or treble their size after first reproduction. If individuals of these species are caught when they are comparatively small, the benefit of the additional weight increment from further growth is lost. The catching of too many relatively small fish will

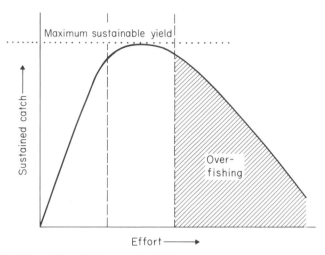

Fig. 12.1 The relationship between the fishing effort and the sustainable catch of fish.

decrease the average size of the individuals in the population and lead to decreased future yields: this is known as 'growth overfishing'. Neither does one want to wait until the individual fish are very large because by then their numbers will have declined as a result of natural mortality. The compromise can be effected by regulating appropriately the size of the meshes in the nets used to make the catch. Other fish, for example the herring, do not grow much after attaining maturity and here, as indeed in those species in which growth does continue, by catching too many adult fish in total, too many pre-reproductive fish, or too many fish that have only spawned a few times, one can, as suggested above, reduce recruitment in succeeding years and cause 'recruitment overfishing'—a much more serious problem.

Most existing fisheries are already on the overfishing part of the curve in Fig. 12.1 (Fig. 12.2; Table 12.3) as evidenced by declining catches per unit effort (see pp. 213–14 and Fig 8.10). In theory, the remedy is simple. Figure 12.3 shows the dependence of yield (measured as yield per recruit to the fishery) on fishing mortality in the stock of plaice in the southern North Sea. The vertical line, corresponding to a fishing mortality of 0.73, indicates the mortality caused by the fishery in the 1930s. An ultimate gain of some 20% in total catch and a threefold increase in catch per unit effort could have been achieved by reducing the fishing mortality and letting the stock recover. In practice, the cost of this simple remedy would have been two-thirds of the fisherman concerned being out of work and a short-term lack of fish whilst

Table 12.3 The extent of exploitation of some commercially important fish stocks in the north-east Atlantic. (From data in Cushing 1980.)

	Cod (*Gadus morhua*)	Haddock (*Melanogrammus aeglefinus*)	Whiting (*Merlangius merlangus*)	Saithe (*Pollachius virens*)	Hake (*Merluccius merluccius*)	Plaice (*Pleuronectes platessa*)	Sole (*Solea solea*)	Herring (*Clupea harengus*)	Mackerel (*Scomber scombrus*)	Redfish (*Sebastes marinus*)
Barents Sea	+ + + +	+ + + +		+ + + +						+ + +
Norwegian Sea	+ + + + +							+ + + + +		+ + +
Greenland	+	+ + + +								+ + +
Iceland	+ + + +	+ + + +		+ + +						
Faroes	+ + + +	+ + + +		+						
North Sea	+ + + +	+ + + +	+ + + +	+ + + + +		+ + +	+ + + +	+ + + + +	+ + + +	
Celtic Sea								+ + +		
Irish Sea	+ + + +		+ + + +			+ + + + +	+ + +	+		
West Scotland	+ + + +	+ + +		+ + +				+ + + + +		
Biscay–Shetland					+ + + +				+	

+ + + + + Catches banned because of overexploitation.
+ + + + Stocks overexploited.
+ + + Stocks exploited at MSY.
+ Stocks underexploited.

295

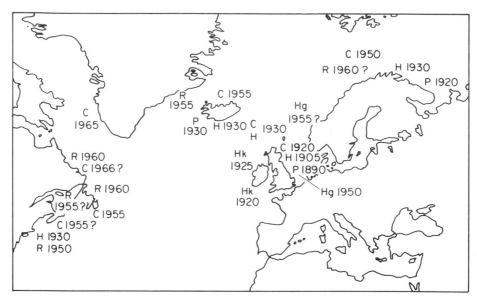

Fig. 12.2 Overfishing in the North Atlantic Ocean. The years are those by which increased fishing effort on the stocks indicated yielded no sustained increase in the catch. C = cod (*Gadus morhua*); H = haddock (*Melanogrammus aeglefinus*); P = plaice (*Pleuronectes platessa*); R = redfish (*Sebastes marinus*); Hk = hake (*Merluccius merluccius*); and Hg = herring (*Clupea harengus*). (After FAO 1968.)

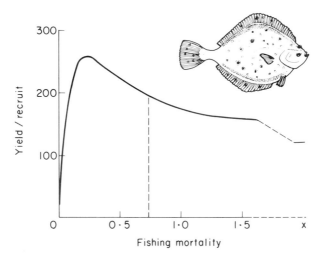

Fig. 12.3 The relationship between yield (measured as yield per recruit to the stock) and fishing mortality for the plaice, *Pleuronectes platessa*, of the southern North Sea. The line at a mortality of 0.73 indicates the position during the 1930s. (After Beverton & Holt 1957.)

the stock recovered. In other words, solutions to overfishing problems have sociological, economic and political repercussions. What in fact was done was to ease that part of the decline that was due to growth overfishing by increasing the mesh size of the nets, so that the same numbers of fisherman could be employed, but thereafter catching fewer, though larger, fish.

The international haggling and acrimony over fishing restrictions, allocations of quotas, etc. will be all too familiar to the reader from the news media. As familiar may also be the process whereby fisheries scientists calculate yields which should not be exceeded and various governments then multiply these estimates by various factors for socioeconomic reasons, or even, in some cases, completely ignore them by failing to ratify the 'agreements' concerned. The decline of the larger baleen whales, first in the Arctic and then in the Antarctic (Fig. 12.4), exemplifies this pattern and also demonstrates the effects which removal of the dominant nektonic species can have on the system of which they were part.

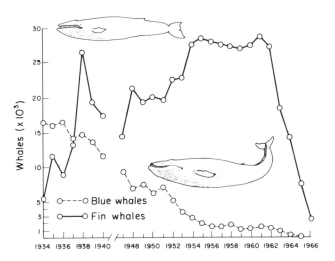

Fig. 12.4 The catch of blue and fin whales (*Sibbaldus musculus* and *Balaenoptera physalus*) in the Antarctic, 1934–1966. (After Small 1971.)

This is not the place to review the series of failures to adopt means whereby the stocks of, for example, the blue whale, *Sibbaldus musculus*, could be conserved (see Small (1971) for such an account), but we can note here all the classic symptoms of overfishing. Catch per unit effort declined from 1936 onwards (Fig. 12.5), yet high catches were still maintained through to 1961

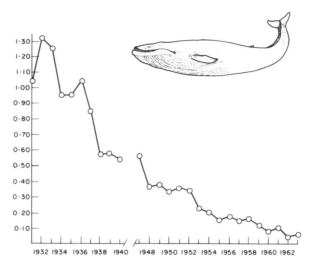

Fig. 12.5 The catch per unit effort (measured as the catch per catcher-day's work) of blue whales (*Sibbaldus musculus*) in the Antarctic, 1931–1963. (After Small 1971.)

(Fig. 12.4) as a result of a greatly increased whaling effort which exacerbated both growth overfishing (Fig. 12.6) and recruitment overfishing (Fig. 12.7). By 1963, 80% of the blue whales caught were sexually immature. Finally, when no more could be caught, the species was protected.

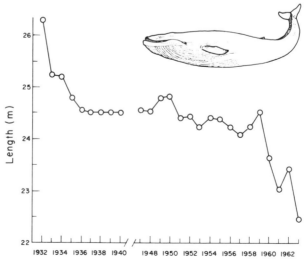

Fig. 12.6 The average length of female blue whales (*Sibbaldus musculus*) of over 21 m in length killed in the Antarctic, 1931–1963. (After Small 1971.)

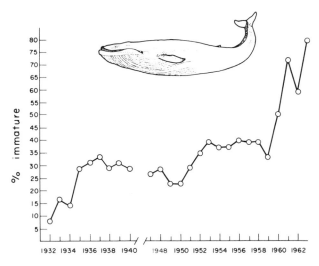

Fig. 12.7 The proportion of sexually-immature individuals of both sexes in the Antarctic blue whale (*Sibbaldus musculus*) catch, 1931–1963. (After Small 1971.)

The decline of the blue and other baleen whales resulted in a decreased level of predation on the krill. This reduction in interspecific competition for food appears to have permitted increased population densities of the other krill-feeders (May *et al.* 1979), including that of the crab-eater seal, *Lobodon carcinophagus*, which has comb-like cusps on its teeth for filtering these crustaceans from the water. Populations of this species have increased dramatically over the last 20 years and it is now by far the most abundant pinniped in the world. As mentioned above, man is now taking an interest in the krill. Exploitation has already started and if various technological problems can be overcome and the estimates of yield are correct, the yield may eventually equal that from all the fish put together. On past showing, however, it is entirely possible that the krill may be decimated before adequate measures are taken to regulate the fishery.

12.2 Pollution

An incredible array of objects and substances are discharged deliberately or accidentally into the sea: mining spoil, oil and petroleum products, industrial solvents, insecticides and herbicides, poison-gas containers, heavy metals, domestic 'rubbish', radionuclides, nuclear wastes, sewage and other organic materials, automobile bodies, industrial cooling water, and a host of others. Most affected are the coastal zones into which many of the above are pumped, tipped or jettisoned, but even in mid ocean pollution is evident—Heyerdahl recorded many overt 'incidents' during his transoceanic voyages. Among coastal areas, pollution is most severe in estuaries and semi-enclosed bays or harbours; in fact in any water body which has a slow rate of exchange in relation to its volume. Thus it is especially noticeable in landlocked seas such as the Baltic and Mediterranean which are separated from the adjacent seas or oceans by shallows. Public bathing is now prohibited in a number of Mediterranean areas because of the risk to human health; and as a result of discharges from the surrounding shores, the oxygen content of the bottom water in the Baltic Sea has decreased from some 20–40% saturation before 1910 to less than 10% after 1960 (Fig. 12.8).

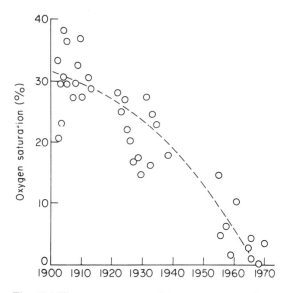

Fig. 12.8 The oxygen content of the bottom water of the Baltic Sea, 1900–1970. (After Jansson 1972.)

Examples could be given almost endlessly: oil spills at Santa Barbara, California, and in the western English Channel; chronic oil pollution of salt-marshes near refineries; high concentrations of heavy metals, PCBs (polychlorinated biphenyls) and pesticides in the tissues of organisms, often concentrated by factors of up to 2.5×10^6 times in the tissues of the top predators and so on. However, the pages of any issue of the *Marine Pollution Bulletin* may be consulted for up-to-the-minute accounts, and this brief section will concentrate on the effects of pollution on marine communities and on how it may be diagnosed, with special reference to organic pollution.

In severe cases, there is no difficulty in detecting the presence of polluting substances or in observing their effect. Sea-birds coated in oil, for example, and extensive areas of deoxygenation are all too obvious. Anoxia is most often associated with inputs of sewage and other organic materials (paper-mill effluent, etc.) The extent to which a given discharge will deplete oxygen in the receiving area is usually measured as the five-day Biochemical Oxygen Demand (BOD): the quantity of oxygen used in five days for the partial oxidation of a sample of the effluent under standard conditions. For sewage, the five-day BOD represents about 70% of the total oxygen demand. The crude sewage from one million people is discharged into the Mersey Estuary and that from one further million is shared between the Tees and the Humber Estuaries, England. This, and a number of other discharges, creates a BOD in the Tees of 253 000 kg per day and in the Mersey of 219 000 kg per day. The total daily oxygen demand of materials voided to the Tees, for example, is therefore equivalent to the removal of all the oxygen from a volume of over 25×10^6 m^3 of water. The effect of such oxygen demands is evident in Fig. 12.9.

In regions of maximal organic matter and minimal oxygen concentration, the biota—other than bacteria—is reduced to a few

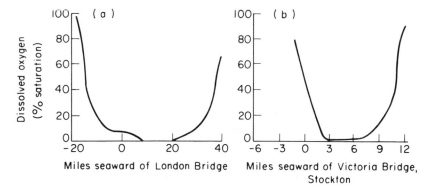

Fig. 12.9 Deficits in the dissolved oxygen content of the water in two estuaries as a result of pollution. (a) The average value for 1950–1959 (third quarter of each year) for the Thames, UK; (b) the Tees, UK, June 1967. (After Barnes 1974.)

highly opportunistic annelids (particularly the polychaetes *Capitella* and *Heteromastus*, and several oligochaete species collectively known as sludge worms). These have been suggested to serve as good indicators of organic pollution, but they also occur in a variety of non-polluted habitats as well. Indeed, they are classic *r*-selected opportunists and *Capitella* occurs, often in large numbers, whenever and wherever an opportunity presents itself (see p. 244). *Capitella*, although not a particularly tolerant genus, has a very short life span and reproduces copiously throughout the year, producing both pelagic and benthic larvae as the occasion demands (see also Chapter 9). It can soon invade and build up a high population density when other species have been eliminated from the benthos by any form of disturbance, e.g. dredging, dinoflagellate blooms (p. 94), storms, oil pollution, etc., as well as by an influx of oxygen-consuming organics. If disturbance is continual, *Capitella* and the other species with a similar strategy may permanently occupy or unceasingly recolonize the habitat.

Deoxygenation of the water is caused by bacterial oxidation of the organic effluent and in order to reduce the oxygen demand of such a discharge, treatment plants are often established. In these, oxidation takes place in filter beds, yielding inorganic nitrates and phosphates, which are discharged to the sea instead of the crude organics. The installation of these plants has resulted in a marked improvement in the oxygen status of several water masses. Each human, however, excretes an average of 9 g nitrogen and 2 g phosphorus per day, so that one million people will contribute 9000 kg and 2000 kg, respectively, of these nutrient elements to the receiving water each day. Further quantities of phosphate will be discharged through sewers in the form of detergents, and of both nutrients via run-off from agricultural land. Hence, although nitrates and phosphates do not exert a BOD, the magnitude of this nutrient injection into a system hitherto receiving unoxidized sewage may be quite sufficient to cause eutrophication. Massive blooms of phytoplankton and of benthic algae can be triggered (especially of macroscopic, annual, green algae) and on the eventual decomposition of this production, deoxygenation can once again result.

These examples have been of circumstances in which the pollutional load is overt and the response of the organisms is evident, most usually as a dramatic reduction in numbers of species. In other cases, the polluting substance may be more insidious or one may wish to detect—and endeavour to prevent—a source of pollution before its affects achieve a level at which they are immediately obvious (Royal Society of London 1979). Gray (1979)

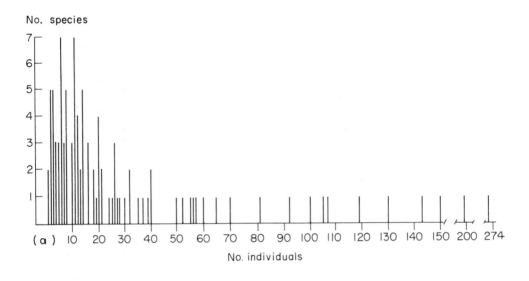

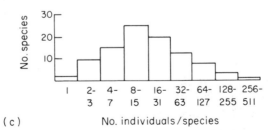

Fig. 12.10 A sample from a hypothetical system comprising 100 species and 3066 individuals arranged to display: (a) the numbers of species represented by differing numbers of individuals; (b) the numbers of species represented by differing size classes of numbers of individuals in a geometric (log₂)scale; and (c) the lognormal distribution which results when (b) above is displayed in histogram form. See also Fig. 12.11.

has suggested that the pattern of the relative abundance of individuals within different species can provide a sensitive means of detecting low levels or the early stages of pollution. In many communities, whether pelagic or benthic, the number of species represented by different numbers of individuals is distributed lognormally; i.e. if one plots the numbers of species which are represented in the community by different numbers of individuals arranged in a series of geometric size categories (a $\log_2$ or $\log_3$ series of size categories is often used) a normal curve results (see Fig. 12.10). One of the useful properties of such a normal curve is that, if, instead of plotting *numbers* of species one plots values of *cumulative percentage* of the total numbers of species present, a straight line is found if the relationship is presented on probability graph paper (Fig. 12.11). (The same property is utilized in the analysis of the particle-size composition of benthic sediments.)

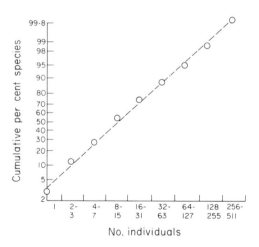

Fig. 12.11 The data of Fig. 12.10 plotted on probability paper as cumulative per cent species against the size classes of numbers of individuals. The straight-line relationship indicates a lognormal distribution.

The early stages of several forms of pollutional disturbance are manifested by breaks in this lognormal line and by an increase in the numbers of the geometric size classes required to cater for the numbers of individuals per species (see Fig. 12.12 which records the changes in a benthic community consequent on the discharge of organic pollutants). Eventually, a straight-line relationship may again be established, but one of shallower slope, spanning many more geometric size classes (Fig. 12.12). The sudden break in the straight line (e.g. 1967 and 1968 in Fig. 12.12) probably simply indicates a disturbance to the community, and also occurs when marine systems are perturbed naturally. Under slight loading of an organic pollutant, for example, the break is caused by a

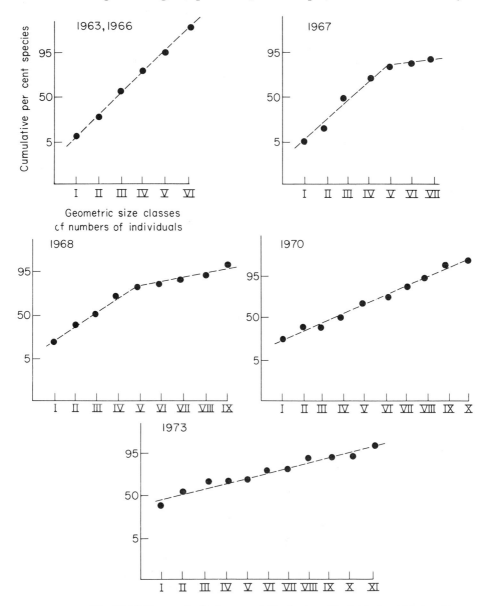

Fig. 12.12 Changes in the lognormal distribution of individuals within benthic species (see Figs. 12.10 and 12.11) as a consequence of the pollution of a sea loch. (After Gray's 1981 analysis of the data of Pearson 1975.)

few species becoming more abundant whilst little change is effected in the others. As loading increases, a few species become superabundant and many others are eliminated, creating both a shallower-sloped and longer line. The importance of this technique is that these changes can be detected long before they could be by other means, for example by monitoring species diversity.

It is important to note that different species are differentially susceptible to almost any pollutant (Table 12.4) and that their response will vary seasonally, with stage in the life cycle (including reproductive status) and with the nature of the environment. A given species may therefore appear tolerant under one set of conditions but susceptible under another: this makes interpretation of the results of standard laboratory 'toxicity-

Table12.4 The toxicity of various pollutants of water to the main groups of aquatic animals: the greatest number of plus symbols indicates the highest toxicity (especially sensitive or tolerant species in each group have been ignored). (After Nelson-Smith 1972.)

Pollutant	Holoplankton and meroplankton	Miscellaneous benthic invertebrates	Crustaceans	Molluscs	Fish
Heavy metal salts:					
copper	+ + +		+ + + +	+ + + +	+ + + +
lead			+ +		+ + + +
zinc	+ +		+ + +	+ + +	+ + +
mercury	+ + + + +		+ + + +	+ + + +	+ + + +
cadmium					+ + + + +
Chlorine		+ + +	+ + + +	+ + +	+ + + +
CNCl					+ + + + +
Cyanide			+ + + +	+ + +	+ + + + +
Fluoride					+ + +
Sulphide					+ + + +
Mercaptan					+ + + +
Phenol	+ + + +		+ + +	+ +	+ + +
Cresol			+ + +	+ + +	+ + +
Formaldehyde			+ +		+ +
Herbicides:					
paraquat; simazine			+ + +		+ + + +
pentachlorophenate		+ + + +	+ + + +	+ + + +	+ + + +
2,4-D					+ +
Pesticides:					
rotenone			+ + + + +		+ + + + +
chlorinated HC; PCB			+ + + + +		+ + +
organophosphorus			+ + + + +		+ + + + +
Crude and fuel oils	+ + +	+	+	+	+ +
Low-aromatic HC			+ + +		+ + + +
Light oil products	+ +				+ +
Oil-spill cleansers:					
early (e.g. BP1002)			+ + + +	+ + +	+ + + +
modern (e.g. BP1100)			+ +	+	+
Surfactants	+ + + +	+ + + +	+ + +	+ + +	+ + +

testing' extremely hazardous. Species near the margins of their range and already under some degree of environmental stress are likely to be much more susceptible to a pollutant than they would be under a more favourable environmental regime. The same can apply when under, or free from, pressure from competitors. Pollutants can also act synergistically; i.e. two pollutants which can be withstood when experienced separately have a much greater effect when applied in combination. Thus ecological methods of assessing pollutional changes, which integrate all these variables, are generally preferable to laboratory studies on test organisms.

Finally, as in section 12.1, we must involve socioeconomic factors in the discussion. Pollution is almost always avoidable. Accidental spills could be eliminated by more care or thoughtfulness. Deliberate discharges could be rendered harmless at source—at a price. The operative factors are simply time and, above all, money. If a political party adopted as its platform the ending of pollution, by requiring polluters to denature before discharge or by undertaking these works itself, then either the cost of the operations would be added to the finished products or would be borne by taxation. In either event, pollution would disappear at the price of a significant increase in the cost of living. How many votes would such a platform attract? The choice is ours and yours: a cleaner environment and higher prices; or cheap commodities and the discharges listed at the start of this section.

12.3 Reclamation

In small quantities, some pollutants stimulate marine productivity; reclamation destroys it. 'Reclamation' is hardly the most appropriate term for the process concerned. One reclaims that which one has lost, although much reclamation from the sea is more in the nature of a pre-emptive strike than of a just attempt to recover something lost, stolen or strayed. The word, however, has decades of use in its pedigree and is apparently the only one available.

Reclamation is a feature of the immediately coastal fringe of the sea and, so far, has been concentrated in estuaries, lagoons, bays and intertidal expanses of sediment, salt-marsh or mangrove-swamp. Historically, dating back to the Romans, most reclamation was for the purpose of either or both flood protection and the production of agriculturally valuable land. In more modern times it has also been used to create land for residential housing, industry (electricity-generating stations, petrochemical and oil refinery installations, etc.) and port, dock and airport facilities, as well as to yield empoundments suitable as freshwater storage reservoirs and to form systems for the generation of electricity by hydroelectrical means. Whole new residential/industrial/transportation complexes have been built on land reclaimed for the purpose.

A brief review of land reclamation in the Netherlands—where much of the technology was pioneered and where the most ambitious reclamation schemes have been carried out—will serve to illustrate the magnitude of the phenomenon, although the gaining of land from the sea is now world-wide. Land reclamation was already in full swing in The Netherlands in the twelfth century and it has increased in scale since then, such that now 18% of the area of that country is reclaimed land (Fig. 12.13). Besides a number of relatively small operations, two large projects have been undertaken this century—affecting the Zuiderzee (Fig. 12.14) and the 'Delta Region' of the Rhine–Maas Estuary (Fig. 12.15)—whilst a third, involving much of the Waddenzee, is under debate.

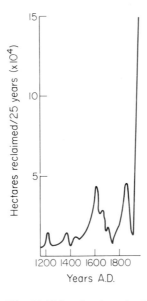

Fig. 12.13 Land reclamation in The Netherlands from the thirteenth century to the present day. (From data in Wagret 1959.)

The 350 000 ha Zuiderzee was enclosed by building a 32 km long barrage across its mouth, with walls of boulder clay dredged from the Zuiderzee itself and with a 30 m thick infill of sand taken from the Waddenzee. Final separation from the North Sea

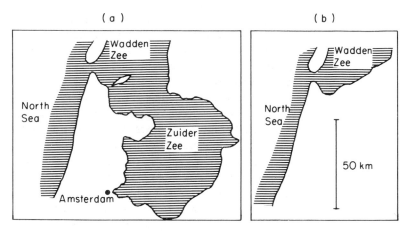

Fig. 12.14 The Zuider Zee: (a) before reclamation; (b) after reclamation. Marine regions are shaded.

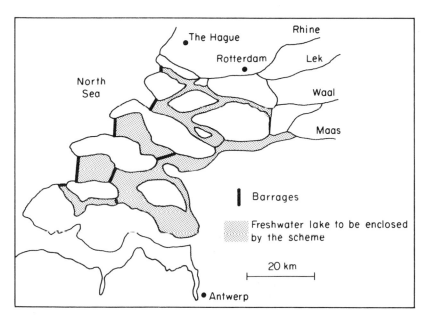

Fig. 12.15 The 'Delta Plan' for the reclamation of the Rhine–Maas Estuary as originally conceived. (After Barnes 1977b.)

and Waddenzee was effected in 1932 and a large lake, the IJssel-meer, was thereby formed. The waters of the IJsselmeer grad-ually became fresh as river input flushed out the salts through sluices which permitted outflow but not sea-water inflow.

309

Creation of this lake was only the first phase of the scheme. There remained the task of reclaiming 220 000 ha of agricultural land in five large 'polders' (four of which have been completed) from regions of silt clay sediments within the erstwhile marine bay. Reclamation proceeded by isolating each proposed region of land with further barrages and then by pumping out the empounded water to produce a land surface as much as 5 m below sea-level. After this surface had dried, whole agricultural communities were established.

The 'Delta Plan' for the reclamation of the Rhine–Maas Estuary, on which work started after the disastrous storm surge in the North Sea in the winter of 1953, is broadly similar, although much of that area—including the Europort at Rotterdam—was already reclaimed land. Originally, the overall object was the construction of a series of barrages across all the estuarine mouths except that of the Westerschelde, thereby considerably shortening the length of the sea walls required for coastal protection, and creating another large freshwater system from which 16 000 ha of potentially fertile land could be reclaimed. In fact, the Grevelingen Estuary, although dammed, may remain as a saline lake, and conservation pressure has permitted the Oosterschelde to remain partly tidal.

The Zuiderzee marine fisheries are now clearly extinct and the effects of the Delta Plan and the proposals for reclamation of the Waddenzee on the remaining fisheries can be gauged from Fig. 12.16. Both areas are the nursery grounds of flat-fish (see pp. 204–11). The Waddenzee nursery may account for one-third of the sole and plaice later caught, as adults, in the North Sea; it certainly holds some 65% of all young sole and 80% of all young plaice in Dutch coastal waters, and it is also used by several other fish of commercial importance, including herring. By and large, these figures speak for themselves.

The scale of these Dutch examples is dramatic, but many other schemes have been completed or are proposed for the bays and estuaries of the north temperate zone. From the standpoint of conservation of these areas, the problem has been the piecemeal nature of reclamation and the lack of overall plans. Good local arguments may be adduced, in isolation, for the reclamation of a particular site, and one only finds out later that the number of such sites in total is large. Reclamation of one area may have repercussions on adjacent sites: it may be argued, for example, that species displaced from area A can be accommodated in area B, whilst at the same time it is being argued further along the coast that species displaced from B can be accommodated at A!

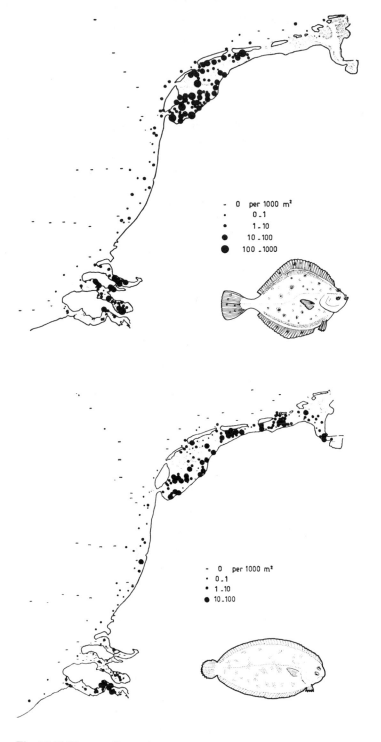

Fig. 12.16 The use of coastal areas of The Netherlands being reclaimed or threatened with reclamation in the future as nursery grounds for: (a) plaice (*Pleuronectes platessa*); and (b) sole (*Solea solea*). Data as numbers of fish under 13 cm length per 1000 m^{-2} in April 1969 (plaice) or May 1970 (sole). (After Zijlstra 1972.)

What effects wholesale reclamation would have are largely unknown. Most studies have concerned only single sites. Attention has also largely been focussed only on the top consumers—the waders, wildfowl and commercial fish species. Since many of the fisheries are in decline anyway, the advantage of reclamation are seen as outweighing the disadvantage to the fishery (especially as, in many cases, one nation's nursery area provides another nation's fishery). In respect of the displaced birds, it is relatively easy to construct and manage reserves for this element of the fauna, and there have been many notable successes: including several associated with the Dutch schemes mentioned above.

Birds and fish are attracted to such shallow coastal systems because they are especially productive (pp. 38–40), however, and these vertebrates only represent the visible tip of the iceberg. It is far more difficult to conserve, or even to quantify, the importance of these areas as centres of high productivity. Microalgal productivity and detritus production do not have the popular appeal of geese, curlew and avocets. Yet not only are coastal birds (for all their splendour) amongst the most mobile of coastal organisms, they are also amongst the least important members of the food web and production process. Unfortunately, not until we have answers to many of the questions raised in this book on the precise significance of the littoral zone in relation to the productivity of the whole marine system will we be in a position to argue for the conservation of coastal wetlands on a firm scientific basis.

References

ACHITUV Y. & BARNES H. (1978) Some observations on *Tetraclita squamosa rufotincta* Pilsbry. *J. exp. mar. Biol. Ecol.*, **31**, 315–24.

BAINBRIDGE R. (1953) Studies on the inter-relationships of zooplankton and phytoplankton. *J. mar. biol. Ass. U.K.*, **32**, 385–447.

BAK R.P.M. & ENGEL M.S. (1979) Distribution, abundance and survival of juvenile corals (Scleractinia) and the importance of life history strategies in the parental coral community. *Mar. Biol.*, **54**, 341–52.

BARNES D. J. (1973) Growth in colonial scleractinians. *Bull. mar. Sci.*, **23**, 280–98.

BARNES R.S.K. (1974) *Estuarine Biology*. Edward Arnold, London.

BARNES R.S.K. (ed.) (1977a) *The Coastline*. John Wiley & Sons, London.

BARNES R.S.K. (1977b) Introduction: the coastline. In Barnes R.S.K. (ed.), *The Coastline*, pp. 3–27. John Wiley & Sons, London.

BARNES R.S.K. (1980a) *Coastal Lagoons*. Cambridge University Press, Cambridge.

BARNES R.S.K. (1980b) The unity and diversity of aquatic systems. In Barnes R.S.K. & Mann K.H. (eds), *Fundamentals of Aquatic Ecosystems*, pp. 5–23. Blackwell Scientific Publications, Oxford.

BARNES R.S.K. & MANN K.H. (eds) (1980) *Fundamentals of Aquatic Ecosystems*. Blackwell Scientific Publications, Oxford.

BATTAGLIA B. & BEARDMORE J.A. (eds) (1978) *Marine Organisms: Genetics, Ecology and Evolution*. Plenum Press, New York.

BATTAGLIA B., BISOL P.M. & FAVA G. (1978) Genetic variability in relation to the environment in some marine invertebrates. In Battaglia B. & Beardmore J.A. (eds), *Marine Organisms: Genetics, Ecology and Evolution*, pp. 53–70. Plenum Press, New York.

BEHRENS-YAMADA S. (1977) Geographic range limitation of the intertidal gastropods *Littorina sitkana* and *L. planaxis*. *Mar. Biol.*, **39**, 61–5.

BELYAEV G.M. (1966) Bottom fauna of the ultra-abyssal depths of the world ocean (in Russian). *Akad. Nauk. S.S.S.R.*, **591**(9), 1–248.

BEVERTON R.J.H. & HOLT S.J. (1957) On the dynamics of exploited fish populations. *Fish. Invest. (Lond.)*, (Ser. 2), **19**, 1–533.

BOLD H.C. & WYNNE M.J. (1979) *Introduction to the Algae*. Prentice-Hall, New Jersey.

BOWERBANK J.S. (1874) *A Monograph of British Spongiidae*, vol. 3. Ray Society, London.

BRANCH G.M. (1975) Intraspecific competition in *Patella cochlear* Born. *J. Anim. Ecol.*, **44**, 263–82.

BRIGGS J.C. (1974) *Marine Zoogeography*. McGraw-Hill Book Co., London.

BUCHANAN J.B., SHEADER M. & KINGSTON P.R. (1978) Sources of variability in the benthic macrofauna off the south Northumberland coast, 1971–6. *J. mar. biol. Ass. U.K.*, **58**, 191–210.

BUSS L.W. & JACKSON J.B.C. (1979) Competitive networks: nontransitive competitive relationships in cryptic coral reef environments. *Am. Nat.*, **113**, 223–34.

CHAPMAN V.J. (ed.) (1977) *Wet Coastal Ecosystems*. Elsevier, Amsterdam.

CLARK R.B. (1964) *Dynamics in Metazoan Evolution*. Clarendon Press, Oxford.

CONNELL J.H. (1972) Community interactions on marine rocky intertidal shores. *Ann. Rev. Ecol. Syst.*, **3**, 169–92.

CRISP D.J. (1962) *Grazing in Terrestrial and Marine Environments*. (Brit. Ecol. Soc. Symp. 4) Blackwell Scientific Publications, Oxford.

CRISP D.J. (1976) The role of the pelagic larva. In Spencer Davies P. (ed.), *Perspectives in Experimental Zoology*, pp. 145–55. Pergamon Press, Oxford.

CRISP D.J. (1978) Genetic consequences of different reproductive strategies in marine invertebrates. In Battaglia B. & Beardmore J.A. (eds), *Marine Organisms: Genetics, Ecology and Evolution*, pp. 257–73. Plenum Press, New York.

CUNDELL A.M., BROWN M.S., STANFORD R. & MITCHELL R. (1979) Microbial degradation of *Rhizophora mangle* leaves immersed in the sea. *Estuar. coast. mar. Sci.*, **9**, 281–6.

References

CUSHING D.H. (1959) The seasonal variation in oceanic production as a problem in population dynamics. *J. Cons.*, **24**, 455–64.

CUSHING D.H. (1971) Upwelling and the production of fish. *Adv. mar. Biol.*, **9**, 255–334.

CUSHING D.H. (1973) *Recruitment and Parent Stock in Fishes.* (Washington Sea Grant Publication, 73–1.) University of Washington, Seattle.

CUSHING D.H. (1975) *Marine Ecology and Fisheries.* Cambridge University Press, Cambridge.

CUSHING D.H. (1980) European fisheries. *Mar. Poll. Bull.*, **11**, 311–15.

CUSHING D.H. & HARRIS J.G.K. (1973) Stock and recruitment and the problem of density dependence. *Rapp. Proc. Cons. perm. int. Explor. Mer*, **164**, 142–55.

CUSHING D.H. & WALSH J.J. (eds) (1976) *The Ecology of the Seas.* Blackwell Scientific Publications, Oxford.

DALY J.M. (1972) The maturation and breeding biology of *Harmothoë imbricata* (Polychaeta: Polynoidae). *Mar. Biol.*, **12**, 53–66.

DAVID P.M. (1965) The surface fauna of the ocean. *Endeavour*, **24**, 95–100.

DAY J.W., SMITH W.G., WAGNER P.R. & STOWE W.C. (1973) Community structure and carbon budget of a salt-marsh and shallow bay estuarine system in Louisiana. *Center Wetland, Louisiana State Univ. Publ.*, LSU-SG-72-04.

DAYTON P.K. (1971) Competition, disturbance, and community organization: the provision and subsequent utilization of space in a rocky intertidal community. *Ecol. Monogr.*, **41**, 351–89.

DAYTON P.K. (1975) Experimental studies of algal canopy interactions in a sea otter-dominated kelp community at Amchitka Island, Alaska. *Fish. Bull.*, **73**, 230–7.

DAYTON P.K. & HESSLER R.R. (1972) Role of biological disturbance in maintaining diversity in the deep sea. *Deep Sea Res.*, **19**, 199–208.

DE BEER G. (1958) *Embryos and Ancestors*, 3rd edn. Oxford University Press, Oxford.

DÖRJES J. & HOWARD J.D. (1975) Estuaries of the Georgia coast, USA: Sedimentology and biology. IV Fluvial–marine transition indicators in an estuarine environment, Ogeechee River–Ossabaw Sound. *Senckenb. marit.*, **7**, 137–9.

EKMAN S. (1967) *Zoogeography of the Sea.* Sidgwick & Jackson, London.

ENDEAN R. (1973) Population explosions of *Acanthaster planci* and associated destruction of hermatypic corals in the Indo-West Pacific region. In Jones O.A. & Endean R. (eds), *Biology and Geology of Coral Reefs II: Biology I*, pp. 389–438. Academic Press, New York.

EPPLEY R.W., ROGERS J.N. & McCARTHY J.J. (1969) Half saturation constants for uptake of nitrate and ammonium by marine phytoplankton. *Limnol. Oceanogr.*, **14**, 912–20.

FAO (1968) The state of world fisheries. *World Food Problems*, 7.

FAO (1978) *Yearbook of Fishery Statistics*, vol. 44. FAO, Rome.

FENCHEL T. (1975) Character displacement and coexistence in mud snails (Hydrobiidae). *Oecologia (Berl.)*, **20**, 19–32.

FENCHEL T. (1978) The ecology of micro- and meiobenthos. *Ann. Rev. Ecol. Syst.*, **9**, 99–121.

FENCHEL T. & RIEDEL R.J. (1970) The sulphide system: a new biotic community underneath the oxidised layer of marine sand bottoms. *Mar. Biol.*, **7**, 255–68.

FIELD J.G., JARMAN N.G., DIECKMANN G.S., GRIFFITHS C.L., VELIMEROV B. & ZOUTENDYK P. (1977) Sun, waves, seaweed and lobsters: the dynamics of a west coast kelp-bed. *S. Af. J. Sci.*, **73**, 7–10.

FISHER R.A. (1930) *The Genetical Theory of Natural Selection.* Clarendon Press, Oxford.

FOGG G.E. (1980) Phytoplanktonic primary production. In Barnes R.S.K. & Mann K.H. (eds), *Fundamentals of Aquatic Ecosystems*, pp. 24–45. Blackwell Scientific Publications, Oxford.

FRETTER V. & GRAHAM A. (1980) *The Prosobranch Molluscs of Britain and Denmark. Part 5—Marine Littorinacea.* J. Moll. Stud. Suppl. 7.

References

FRIEDRICH H. (1965) *Meeresbiologie*. Borntraeger, Berlin.

GEORGE J.D. & GEORGE J.J. (1979) *Marine Life*. Harrap, London.

GERLACH S.A. (1978) Food chain relationships in subtidal silty sand marine sediments and the role of meiofauna in stimulating bacterial productivity. *Oecologia (Berl.)*, **33**, 55–69.

GOLDBERG E.G. (1972) *A Guide to Marine Pollution*. Gordon & Breach Science Publications, New York.

GOULD S.J. (1977) *Ontogeny and Phylogeny*. Belknap, Cambridge, Massachusetts.

GRASSLE J.F. & GRASSLE J.P. (1977) Temporal adaptations in sibling species of *Capitella*. In Coull B.C. (ed.), *Ecology of Marine Benthos*, pp. 177–88. University of South Carolina Press, Columbia.

GRAY J.S. (1979) Pollution-induced changes in populations. *Phil. Trans. r. Soc. Lond. B*, **286**, 545–61.

GRAY J.S. (1981) *The Ecology of Marine Sediments: An Introduction to the Structure and Function of Benthic Communities*. Cambridge University Press, Cambridge.

GRAY J. & BOUCOT A.J. (eds) (1979) *Historical Biogeography, Plate Tectonics and the Changing Environment*. Oregon State University Press, Corvallis, Oregon.

GRIFFITHS R.J. (1977) Reproductive cycles in littoral populations of *Choromytilus meridionalis* (Kr.) and *Aulacomya ater* (Molina) with a quantitative assessment of gamete production in the former. *J. exp. mar. Biol. Ecol.*, **30**, 53–71.

GOHAR H.A.F. & SOLIMAN G.N. (1963) On the mytilid species boring in living corals. *Publ. Mar. Biol. Stn. Ghardaqua*, **12**, 65–92.

GOREAU T.F. & GOREAU N.I. (1973) The ecology of Jamaican coral reefs. II. Geomorphology, zonation and sedimentary phases. *Bull. mar. Sci.*, **23**, 399–464.

GULLAND J.A. (ed.) (1977) *Fish Population Dynamics*. John Wiley & Sons, London.

HARDEN JONES F.R. (1980) The nekton: production and migration patterns. In Barnes R.S.K. & Mann K.H. (eds), *Fundamentals of Aquatic Ecosystems*, pp. 119–42. Blackwell Scientific Publications, Oxford.

HARDY A.C. (1924) The herring in relation to its animate environment. Part 1. The food and feeding habits of the herring with special reference to the east coast of England. *Fish. Invest. (Lond.)*, (Ser. 2), 7(3), 1–53.

HARDY A.C. (1962) *The Open Sea: Its Natural History. Part I, The World of Plankton*. Collins, Sons & Co., London.

HARPER J.L. (1977) *Population Biology of Plants*. Academic Press, London and New York.

HEEZEN B.C. & HOLLISTER C.D. (1971) *The Face of the Deep*. Oxford University Press, New York.

HEINRICH A.K. (1962) The life histories of plankton animals and seasonal cycles of plankton communities in the oceans. *J. Cons.*, **27**, 15–24.

HERON A.C. (1972) Population ecology of a colonizing species: the pelagic tunicate *Thalia democratica*. 1. Individual growth rate and generation time. *Oecologia (Berl.)*, **10**, 269–93.

HESSLER R.R. & SANDERS H.L. (1967) Faunal diversity in the deep sea. *Deep Sea Res.*, **14**, 65–78.

HEUVELMANS B. (1968) *In the Wake of the Sea-serpents*. Hart-Davis, London.

HIGHSMITH R.C. (1979) Coral growth rates and environmental control of density banding. *J. exp. mar. Biol. Ecol.*, **37**, 105–25.

HIGHSMITH R.C., RIGGS A.C. & D'ATTONIO C.M. (1980) Survival of hurricane-generated coral fragments and a disturbance model of reef calcification/growth rates. *Oecologia (Berl.)*, **46**, 322–9.

HOLLAND A.F., MOUNTFORD N.K., HIEGEL M.H., KAUMEYER K.R. & MIHURSKY J.A. (1980) Influence of predation on infaunal abundance in Upper Chesapeake Bay, USA. *Mar. Biol.*, **57**, 221–35.

HOOD D.W. (ed.) (1971) *Impingement of Man on the Oceans*. Wiley-Interscience, New York.

315

References

HOUSE M.R. (ed.) (1979) *The Origin of Major Invertebrate Groups.* Academic Press, London.

HUGHES R.N. (1977) The biota of reef-flats and limestone cliffs near Jeddah, Saudi Arabia. *J. nat. Hist.,* **11**, 77–96.

HUGHES R.N. (1980a) Predation and community structure. In Price J.H., Irvine D.E.G., & Farnham W.F. (eds), *The Shore Environment, Vol. 2: Ecosystems,* pp. 699–728. Syst. Ass. Spec. Vol. 17 (6).

HUGHES R.N. (1980b) Strategies for survival of aquatic organisms. In Barnes R.S.K. & Mann K.H. (eds), *Fundamentals of Aquatic Ecosystems,* pp. 162–84. Blackwell Scientific Publications, Oxford.

HUGHES R.N. & ROBERTS D.J. (1980) Reproductive effort of winkles (*Littorina* spp.) with contrasted methods of reproduction. *Oecologia (Berl.),* **47**, 130–6.

HUGHES R.N. & ROBERTS D.J. (1981) Comparative demography of *Littorina rudis, L. nigrolineata* and *L. neritoides* on three contrasted shores in North Wales. *J. Anim. Ecol.,* **50**, 251–68.

HUSTON M. (1979) A general hypothesis of species diversity. *Am. Nat.,* **113**, 81–101.

HUTCHINSON G.E. (1967) *A Treatise on Limnology. Vol. II. Introduction to Lake Biology and the Limnoplankton.* John Wiley & Sons, New York.

JACKSON J.B.C. (1979) Morphological strategies of sessile animals. In Larwood G. & Rosen B.R. (eds), *Biology and Systematics of Colonial Organisms,* pp. 499–555. Syst. Ass. Special Vol. 11.

JANSSON B.O. (1972) *Ecosystem Approach to the Baltic Problem.* Swedish National Science Research Council, Stockholm.

JOHNSON W.S., GIGON A., GULMON S.L. & MOONEY H.A. (1974) Comparative photosynthetic capacities of intertidal algae under exposed and submerged conditions. *Ecology,* **55**, 450–3.

JOHNSTONE J., SCOTT A. & CHADWICK H.C. (1924) *The Marine Plankton.* Hodder & Stoughton, London.

JONES O.A. & ENDEAN R. (1973) *Biology and Geology of Coral Reefs. II: Biology 1.* Academic Press, New York.

JØRGENSEN C.B. (1976) August Pütter, August Krogh, and modern ideas on the use of dissolved organic matter in aquatic environments. *Biol. Rev.,* **51**, 291–328.

KETCHUM B.H. (ed.) (1972) *The Water's Edge.* MIT Press, Cambridge, Massachusetts.

KNIGHTS B. & PHILIPS A.J. (eds) (1979) *Estuarine and Coastal Land Reclamation and Water Storage.* Saxon House, Farnborough.

KOBLENTZ-MISHE O.J., VOLKOVISNY V.V. & KABANOVA J.G. (1970) Plankton primary production of the world ocean. In Wooster W.S. (ed.), *Scientific Exploration of the South Pacific,* pp. 183–93. National Academy of Science, Washington.

KOHN A.J. (1959) The ecology of *Conus* in Hawaii. *Ecol. Monogr.,* **29**, 47 –90.

KOHN A.J. (1979) Ecological shift and release in an isolated population: *Conus miliaris* at Easter Island. *Ecol. Monogr.,* **48**, 323–36.

KUIPERS B.R., DE WILDE P.A.W.J. & CREUTZBERG F. (1981) Energy flow in a tidal flat ecosystem. *Mar. Ecol. Progr. Ser.,* **5**, 215–21.

LANG J. (1973) Interspecific aggression by scleractinian corals. 2. Why the race is not only to the swift. *Bull. mar. Sci.,* **23**, 260–79.

LASSEN H.H. & TURANO F.J. (1978) Clinal variation and heterozygote deficit at the Lap-locus in *Mytilus edulis. Mar. Biol.,* **49**, 245–54.

LASSIG B.R. (1977) Communication and coexistence in a coral community. *Mar. Biol.,* **42**, 85–92.

LEVINTON J.S. (1977) Ecology of shallow water, deposit-feeding communities Quisset Harbor, Massachusetts. In Coull B.C. (ed.), *Ecology of Marine Benthos,* pp. 191–227. University of South Carolina Press, Columbia.

LEVINTON J.S. (1979) Deposit-feeders, their resources, and the study of resource

References

limitation. In Livingston R.J. (ed.), *Ecological Processes in Coastal and Marine Systems*, pp. 117–41. Plenum Press, New York.

LEVINTON J.S. & LOPEZ G.R. (1977) A model of renewable resources and limitation of deposit-feeding benthic populations. *Oecologia (Berl.)*, **31**, 177–90.

LEWIS J.R. (1964) *The Ecology of Rocky Shores*. English Universities Press, London.

LITTLER M.M. & LITTLER D.S. (1980) The evolution of thallus form and survival strategies in benthic macroalgae: field and laboratory tests of a functional form model. *Am. Nat.*, **116**, 25–44.

LONGHURST A.R. (ed.) (1981) *Analysis of Marine Ecosystems*. Academic Press, London.

LUBCHENCO J. (1978) Plant species diversity in a marine intertidal community: importance of herbivore food preference and algal competitive abilities. *Am. Nat.*, **112**, 23–39.

LÜNING K. (1979) Growth strategies of three *Laminaria* species (Phaeophycae) inhabiting different depth zones in the sublittoral region of Helgoland (North Sea). *Mar. Ecol. Prog. Ser.*, **1**, 195–207.

MACISAAC J.J. & DUGDALE R.C. (1969) The kinetics of nitrate and ammonia uptake by natural populations of marine phytoplankton. *Deep Sea Res.*, **16**, 45–57.

McLAREN I.A. (1965) Some relationships between temperature and egg size, body size, development rate and fecundity, of the copepod *Pseudocalanus*. *Limnol. Oceanogr.*, **10**, 528–38.

McROY C.P. & HELFFERICH C. (eds) (1977) *Seagrass Ecosystems*. Marcel Dekker, New York.

MANGUM C.P. (1976) Primitive respiratory adaptations. In Newell R.C. (ed.), *Adaptation to Environment*, pp. 191–278. Butterworth & Co., London.

MANN K.H. (1973) Seaweeds: their productivity and strategy for growth. *Science (N.Y.)*, **182**, 975–81.

MARE M.F. (1942) A study of a marine benthic community with special reference to the micro-organisms. *J. mar. biol. Ass. U.K.*, **25**, 517–54.

MARSHALL N.B. (1979) *Development in Deep Sea Biology*. Blandford Press, Poole.

MAY R.M., BEDDINGTON J.R., CLARK C.W., HOLT S.J. & LAWS R.M. (1979) The management of multi-species fisheries. *Science (N.Y.)*, **205**, 267–77.

MENGE B.A. (1976) Organization of the New England rocky intertidal community: role of predation, competition and environmental heterogeneity. *Ecol. Monogr.*, **46**, 355–93.

MENZIES R.J., GEORGE R.Y. & ROWE G.T. (1973) *Abyssal Environment and Ecology of the World Ocean*. John Wiley & Sons, New York.

MERGNER H. (1971) Structure ecology and zonation of Red Sea reefs (in comparison with South Indian and Jamaican reefs). In Stoddart D.R. & Yonge C.M. (eds), *Regional Variation in Indian Ocean Coral Reefs*, pp. 141–61. Symp. Zool. Soc. Lond. 28.

MILLAR R.H. (1970) *British Ascidians*. Academic Press, London.

MILLS E.L. (1969) The community concept in marine zoology, with comments on continua and instability in some marine communities: a review. *J. fish. Res. Bd Canada*, **26**, 1415–28.

MUUS B.J. (1967) The fauna of Danish estuaries and lagoons. *Meddr. Danm. fisk.-og Havunders.* (NS), **5**, 1–316.

NAKAMARA K. & SEKIGUCHI K. (1980) Mating behaviour and oviposition in the pycnogonid *Propellane longiceps. Mar. Ecol. Prog. Ser.*, **2**, 163–8.

NELSON-SMITH A. (1972) *Oil Pollution and Marine Ecology*. Elek, London.

NIXON S.W. (1980) Between coastal marshes and coastal waters—a review of twenty years of speculation and research on the role of salt marshes in estuarine productivity and water chemistry. In Hamilton P. & Macdonald K.B. (eds), *Estuarine and Wetland Processes with Emphasis on Modeling*, pp. 437–525. Plenum Press, New York.

317

References

NIXON S.W., OVIATT C.A. & HALE S.S. (1976) Nitrogen regeneration and the metabolism of coastal marine bottom communities. In Anderson J.M. & Macfadyen A. (eds), *The Role of Terrestrial and Aquatic Organisms in Decomposition Processes*, pp. 269–83. Blackwell Scientific Publications, Oxford.

NORTH W.J. (ed.) (1971) The biology of giant kelp beds (*Macrocystis*) in California. *Beih. Nova Hedwigia*, **32**.

ODUM W.E. (1980) Utilization of aquatic production by man. In Barnes R.S.K. & Mann K.H. (eds), *Fundamentals of Aquatic Ecosystems*, pp. 143–61. Blackwell Scientific Publications, Oxford.

OGDEN J.C. & EBERSOLE J.P. (1981) Scale and community structure of coral reef fishes: a long-term study of a large artificial reef. *Mar. Ecol. Progr. Ser.*, **4**, 97–103.

PAINE R.T. (1966) Food web complexity and species diversity. *Am. Nat.*, **100**, 65–75.

PARSONS T.R., TAKAHASHI M. & HARGRAVE B. (1977) *Biological Oceanographic Processes*, 2nd edn. Pergamon Press, New York.

PETERSON B.J. (1980) Aquatic primary productivity and the $^{14}C-CO_2$ method: a history of the productivity problem. *Ann. Rev. Ecol. Syst.*, **11**, 359–85.

PETERSON C.H. (1979) Predation, competitive exclusion, and diversity in the soft-sediment benthic communities of estuaries and lagoons. In Livingston R.J. (ed), *Ecological Processes in Coastal and Marine Systems*, pp. 233–64. Plenum Press, New York.

PHILLIPS R.C. & McROY C.P. (eds) (1980) *Handbook of Seagrass Biology*. Garland, New York.

PIANKA E.R. (1974) *Evolutionary Ecology*. Harper & Row, London.

PIKE G.C. (1962) Migration and feeding of the gray whale (*Eschrichtius gibbosus*). *J. fish. Res. Bd Canada*, **19**, 815–38.

POMEROY L.R. (1979) Secondary production mechanisms of continental shelf communities. In Livingston R.J. (ed.), *Ecological Processes in Coastal and Marine Systems*, pp. 163–86. Plenum Press, New York.

QASIM S.Z. (1970) Some problems related to the food chain in a tropical estuary. In Steele J.H. (ed.), *Marine food chains*, pp. 46–51. Oliver & Boyd, Edinburgh.

RAMUS J., BEALE S.I. & MAUZERALL D. (1976) Correlation of changes in pigment content with photosynthetic capacity of seaweeds as a function of water depth. *Mar. Biol.*, **37**, 231–8.

RANWELL D.S. (1972) *Ecology of Salt Marshes and Sand Dunes*. Chapman & Hall, London.

RAYMONT J.E.G. (1963) *Plankton and Productivity in the Oceans*. Pergamon Press, Oxford.

REID J.L. (1962) On circulation, phosphate phosphorus content, and zooplankton volumes in the upper part of the Pacific Ocean. *Limnol. Oceanogr.*, **7**, 287–306.

REIMOLD R.J. & QUEEN W.H. (eds) (1974) *Ecology of Halophytes*. Academic Press, New York.

ROBERTSON A.I. (1979) Biology and identification of intertidal Polyzoa. *Oecologia (Berl.)*, **38**, 193–202.

ROYAL SOCIETY OF LONDON (1979) *The Assessment of Sublethal Effects of Pollutants in the Sea*. Royal Society of London, London.

RYLAND J.S. (1962) Biology and identification of intertidal Bryozoa. *Field Stud.*, **1**, 1–19.

RYTHER J.H. (1956) Photosynthesis in the ocean as a function of light intensity. *Limnol. Oceanogr.*, **1**, 61–70.

RYTHER J.H. (1969) Photosynthesis and fish production in the sea. *Science (N.Y.)*, **166**, 72–6.

SALE P.S. (1980) The ecology of fishes on coral reefs. *Oceanogr. mar. Biol. Ann. Rev.*, **18**, 367–421.

SANDERS H.L. (1968) Marine benthic diversity: a comparative study. *Am. Nat.*, **102**, 243–82.

References

SCHÄFER W. (1972) *Ecology and Palaeoecology of Marine Environments.* (Translated by I. Oertel.) Oliver & Boyd, Edinburgh.

SCHEFFER V.B. (1973) The last days of the sea cow. *Smithsonian*, **3** (10), 64–7.

SCHELTEMA R.S. (1978) On the relationship between dispersal of pelagic veliger larvae and the evolution of marine prosobranch gastropods. In Battaglia B. & Beardmore J.A. (eds), *Marine Organisms: Genetics, Ecology and Evolution*, pp. 303–22. Plenum Press, New York.

SCHONBECK M. & NORTON T.A. (1978) Factors controlling the upper limits of fucoid algae on the shore. *J. exp. mar. Biol. Ecol.*, **31**, 303–13.

SCHONBECK M. & NORTON T.A. (1980) Factors controlling the lower limits of fucoid algae on the shore. *J. exp. mar. Biol. Ecol.*, **43**, 131–50.

SCHOPF T.J.M., FISHER J.B. & SMITH III C.A.F. (1978) Is the marine latitudinal diversity gradient merely another example of the species area curve? In Battaglia B. & Beardmore J.A. (eds), *Marine Organisms: Genetics, Ecology and Evolution*, pp. 365–86. Plenum Press, New York.

SCHOPF T.J.M. & GOOCH J.L. (1971) Gene frequencies in a marine ectoproct: a cline in natural populations related to sea temperature. *Evolution*, **25**, 286–9.

SCHUMACHER H. (1976) *Korallenriffe.* BLV Verlagsgesellschaft mbH., Munich.

SEARLES R.B. (1980) The strategy of the red algal life history. *Am. Nat.*, **115**, 113–20.

SEED R. & O'CONNOR R.J. (1981) Community organization in marine algal epifaunas. *Ann. Rev. Ecol. Syst.*, **12**, 49–74.

SHEPPARD C.R.C. (1980) Coral cover, zonation and diversity on reef slopes of Chagos atolls, and population structures of the major species. *Mar. Ecol. Progr. Ser.*, **2**, 193–205.

SHINDO S. (1973) General review of the trawl fishery and demersal fish stocks of the South China Sea. *FAO fish. Tech. Pap.*, **120**.

SHORT F.T. (1980) A simulation model of the seagrass production system. In Phillips R.C. & McRoy C.P. (eds), *Handbook of Seagrass Biology: An Ecosystem Perspective*, pp. 277–95. Garland Press, New York.

SIEBURTH J.M. (1979) *Sea Microbes.* Oxford University Press, New York.

SIMON J.L. & DAUER D.M. (1977) Re-establishment of a benthic community following natural defaunation. In Coull B.C. (ed.), *Ecology of Marine Benthos*, pp. 139–54. University of South Carolina Press, Columbia.

SMALDON G. (1979) *British Coastal Shrimps and Prawns.* Academic Press, London.

SMALL G.L. (1971) *The Blue Whale.* Columbia University Press, New York.

SOROKIN Y.I. (1971) On the role of bacteria in the productivity of tropical oceanic waters. *Int. Rev. ges. Hydrobiol.*, **56**(1), 1–48.

SPENCER-DAVIES P., STODDART D.R. & SIGEE D.C. Reef forms of Addu Atoll, Maldive Islands. In Stoddart D.R. & Yonge C.M. (eds), *Regional Variation in Indian Ocean Coral Reefs*, pp. 217–59. Symp. Zool. Soc. Lond. 28.

STEBBING A.R.D. (1973) Competition for space between the epiphytes of *Fucus serratus. J. mar. biol. Ass. U.K.*, **53**, 247–61.

STEHLI F.G. & WELLS J.W. (1971) Diversity and age patterns in hermatypic corals. *Systematic Zool.*, **20**, 115–26.

STEVENSON T.A. & STEVENSON A. (1972) *Life between Tide Marks on Rocky Shores.* W.H. Freeman & Co., San Francisco.

STEWART W.D.P. (ed.) (1974) *Algal Physiology and Biochemistry.* Blackwell Scientific Publications, Oxford.

STODDART D.R. (1971) Environment and history in Indian Ocean reef morphology. In Stoddart D.R. & Yonge C.M. (eds), *Regional Variation in Indian Ocean Coral Reefs*, pp. 3–38. Symp. Zool. Soc. Lond. 28.

STODDART D.R. & YONGE C.M. (eds) (1971) *Regional Variation in Indian Ocean Coral Reefs.* Symp. zool. Soc. Lond. **28**.

STRONG K.W. & DABORN G.R. (1979) Growth and energy utilization of the intertidal isopod *Idotea baltica* (Pallas) (Crustacea: Isopoda). *J. exp. mar. Biol. Ecol.*, **41**, 101–23.

References

SVERDRUP H.U. (1953) On conditions for the vernal blooming of phytoplankton. *J. Cons.*, **18**, 287–95.

SVERDRUP H.U., JOHNSON M.W. & FLEMING R.H. (1942) *The Oceans.* Prentice-Hall, New York.

TAYLOR J.D. & TAYLOR C.N. (1977) Latitudinal distribution of predatory gastropods on the eastern Atlantic shelf. *J. Biogeogr.*, **4**, 73–81.

TAYLOR W.R. (1964) Light and photosynthesis in intertidal benthic diatoms. *Helgol. wiss. Meeresunters.*, **10**, 29–37.

TEAL J.M. (1980) Primary production of benthic and fringing plant communities. In Barnes R.S.K. & Mann K.H. (eds), *Fundamentals of Aquatic Ecosystems*, pp. 67–83. Blackwell Scientific Publications, Oxford.

THORPE J.P., BEARDMORE J.A. & RYLAND J.S. (1978) Genetic evidence for cryptic speciation in the marine bryozoan *Alcyonidium hirsutum. Mar. Biol.*, **49**, 27–32.

THORSON G. (1957) Bottom communities. In Hedgpeth J.W. (ed.), *Treatise on Marine Ecology and Paleoecology, 1 Ecology*, pp. 461–534. Geological Society of America, New York.

THORSON G. (1971) *Life in the Sea.* Weidenfeld & Nicolson, London.

TODD C.D. & DOYLE R.W. (1981) Reproductive strategies of marine benthic invertebrates: a settlement-time hypothesis. *Mar. Ecol. Progr. Ser.*, **4**, 75–83.

TOWNSEND C.R. & HUGHES R.N. (1981) Maximizing net energy returns from foraging. In Townsend C.R. & Calow P. (eds) *Physiological Ecology: An Evolutionary Approach to Resource Use*, pp. 86–108. Blackwell Scientific Publications, Oxford.

TRANTER D.J. (1976) Herbivore production. In Cushing D.H. & Walsh J.J. (eds), *The Ecology of the Seas*, pp. 186–244. Blackwell Scientific Publications, Oxford.

UNDERWOOD A.J. (1978) An experimental evaluation of competition between three species of intertidal prosobranch gastropods. *Oecologia (Berl.)*, **33**, 185–202.

UNDERWOOD A.J. (1979) The ecology of intertidal gastropods. *Adv. mar. Biol.*, **16**, 111–210.

UNDERWOOD A.J. (1980) The effects of grazing by gastropods and physical factors on the upper limit of distribution of intertidal macroalgae. *Oecologia (Berl.)*, **46**, 201–13.

VALENTINE J.W. (1973) *Evolutionary Paleoecology of the Marine Biosphere.* Prentice-Hall, Englewood Cliffs, New Jersey.

VALENTINE J.W. & MOORES E.M. (1974) Plate tectonics and the history of life in the oceans. *Sci. Am.*, **230**(4), 80–89.

VAN ANDEL T.H. (1979) An eclectic overview of plate tectonics, paleogeography, and paleoecology. In Gray J. & Boucot A.J. (eds), *Historical Biogeography, Plate Tectonics and the Changing Environment*, pp. 9–25. Oregon State University Press, Corvallis, Oregon.

VELIMIROV B., FIELD J.G., GRIFFITHS C.L. and ZOUTENDYK P. (1977) The ecology of kelp bed communities in the Benguela upwelling system. *Helgol. wiss Meeresunters.*, **30**, 495–518.

VINOGRADOV M.E. (1961) Feeding of the deep-sea zooplankton. *Rapp. Proc. Cons. perm. int. Explor. Mer*, **153**, 114–20.

VINOGRADOVA N.G. (1962) Vertical zonation in the distribution of the deep sea benthic fauna in the ocean. *Deep Sea Res.*, **8**, 245–50.

WAGRET P. (1959) *Les Polders.* Dunod, Paris.

WEIGERT R.G. (1979) Ecological processes characteristic of coastal *Spartina* marshes of the south-eastern USA. In Jefferies R.L. & Davy A.J. (eds), *Ecological Processes in Coastal Environments*, pp. 467–90. Blackwell Scientific Publications, Oxford.

WHITTAKER R.H. (1969) New concepts of the kingdoms of organisms. *Science (N.Y.)*, **163**, 150–9.

WHITTINGHAM C.P. (1974) *The Mechanism of Photosynthesis.* Edward Arnold, London.

320

References

WICKLER W. (1968) *Mimicry in Plants and Animals*. World University Library, McGraw-Hill, New York and Toronto.

WIEBE P.H. (1970) Small-scale spatial distribution in oceanic zooplankton. *Limnol. Oceanogr.*, **15**, 205–17.

WILLIAMS G.C. (1977) *Sex and Evolution*. Princeton University Press, Princeton, New Jersey.

WOLCOTT T.G. (1973) Physiological ecology and intertidal zonation in limpets (*Acmaea*): a critical look at 'limiting factors'. *Biol. Bull.*, **145**, 389–422.

WOLFF T. (1977) Diversity and faunal composition of the deep-sea benthos. *Nature (Lond.)*, **267**, 780–5.

WOLFF W.J. (1977) A benthic food budget for the Grevelingen Estuary, The Netherlands, and a consideration of the mechanisms causing high benthic secondary production in estuaries. In Coull B.C. (ed.), *Ecology of Marine Benthos*, pp. 267–80. University of South Carolina Press, Columbia.

WOOD E.J.F., ODUM W.E. & ZIEMEN J.C. (1969) Influence of sea-grasses on the productivity of coastal lagoons. In Castañares A.A. & Phleger F.B. (eds), *Lagunas Costeras: un Simposio*, pp. 495–502. Univ. Nac. Auton. Mexico, Mexico City.

WOODIN S.A. (1974) Polychaete abundance patterns in a marine soft-sediment environment: the importance of biological interactions. *Ecol. Monogr.*, **44**, 171–87.

WOODWELL G.M. (1980) Aquatic systems as part of the biosphere. In Barnes R.S.K. & Mann K.H. (eds), *Fundamentals of Aquatic Ecosystems*, pp. 201–15. Blackwell Scientific Publications, Oxford.

WOODWELL G.M., HOUGHTON R.A., HALL C.A.S., WHITNEY D.E., MOLL R.A. & JUERS D.W. (1979) The Flax Pond ecosystem study: the annual metabolism and nutrient budgets of a salt marsh. In Jefferies R.L. & Davy A.J. (eds), *Ecological Processes in Coastal Environments*, pp. 491–511. Blackwell Scientific Publications, Oxford.

WRIGHT J.B. (ed.) (1977–1978) *Oceanography*. Open University Press, Milton Keynes.

YONGE C.M. (1963) The biology of coral reefs. In Russell F.S. (ed.), *Advances in Marine Biology. Vol. 1, The Biology of Coral Reefs*, pp. 209–60. Academic Press, New York.

ZENKEVITCH L.A. & BIRSTEIN J.A. (1956) Studies of the deep-water fauna and related problems. *Deep-Sea Res.*, **4**, 54–64.

ZIEMAN J.C., THAYER G.W., RUBBLEE M.B. & ZIEMAN R.T. (1979) Production and export of sea-grasses from a tropical bay. In Livingston R.J. (ed.), *Ecological Processes in Coastal and Marine Systems*, pp. 21–33. Plenum Press, New York.

ZIJLSTRA J.J. (1972) On the importance of the Waddenzee as a nursery area in relation to the conservation of the southern North Sea fishery resources. *Symp. zool. Soc. Lond.*, **29**, 233–58.

Index

Index

Index

Index

Index